SURVIVAL CRISIS

PLANET IN PERIL

SURVIVAL CRISIS
PLANET IN PERIL
PEOPLE'S RESPONSE ONLY WAY FORWARD

BHARAT DOGRA

Vitasta
LET KNOWLEDGE SPREAD

Published by
Renu Kaul Verma
Vitasta Publishing Pvt Ltd
2/15, Ansari Road, Daryaganj
New Delhi-110 002
info@vitastapublishing.com

ISBN 978-93-86473-48-6

First Edition 2019
MRP ₹850

Editors: Papri Sri Raman, Veena Batra
Cover and layout by Somesh Kumar Mishra
Printed by Chaman Enterprises

This book is dedicated to my wife
Madhu Dogra who has been a constant
and invaluable support of encouragement
during the four decades of my writings

Men who can graft the trees and make the seed fertile and big can find nc way to let the hungry people eat their produce. Men who have created new fruits in the world cannot create a system whereby their fruits may be eaten.

And the failure hangs over the state like a great sorrow.

—John Steinbeck in the
Grapes of Wrath

Remember His Words

Whenever you are in doubt, or when the self becomes too much with you, apply the following test :
Recall the face of the poorest and the weakest man whom you have seen, and ask yourself if the step you contemplate is going to be of any use to him.
Will he gain anything by it?
Will it restore him.... to control over his own life and destiny?
In other words, will it lead to Swaraj for the hungry and spiritually starving millions?
Then you will find your doubts and.... self melt away.

—*Mahatma Gandhi*

That economics is untrue which ignores or disregards moral values. The extension of the law of non-violence in the domain of economics means nothing less than the introduction of moral values as a factor to be considered in regulating international commerce.

—*Mahatma Gandhi*

What we seek to do is substitute false and non-human economics by true and human. Not killing competition but life giving cooperation is the law of the human being.

—*Mahatma Gandhi*

Contents

Introduction *xiii*

Alarm Bells Ringing **1-24**
Survival Crisis like None Other 3
Development and Core Human Values 9
Urgency of Humankind's Protective Role 12
Of Tipping Points and Planetary Boundaries 16
A Destroyed Environment 20

Tipping Points **25-84**
Risk of Nuclear Weapons Increasing 27
Out of Control Robot Weapons 33
Why Climate Change Needs Attention 37
Rapid Extinction of Species 45
Risks of Hazardous Chemicals 50
GM Crops Extremely Harmful 55
Hazards of Mobiles and Towers 60
Meeting Needs of the Grassroots People 65
Youth Depression 69
Failure in Critical Areas 76
Through the Eyes of Aliens 81

Neglected Priorities **85-98**
Indifferent People 87
Need for an Ethical Value System 92
Prioritise Neglected Objectives 97

The Way Forward **99-120**
Finding Hope amidst Despair 101
A Distress Minimising Model of Development 104
Plan to Reduce GHG Emissions 108
Need for a World Government 112

Social Movements **121-174**
Movements for Justice and Equity 123
Need for Better Linkages and Understanding 127
Combine Global Issues with Local Concerns 132
Peace is Important for All 137
Genuine Spirituality Needed to Counter Hatred 141
Convergence of Social Justice and Spirituality Movement 145
It is all about Four Crucial Relationships 149

Pitfalls Of Progress **175-267**
The Myth and Reality of Progress 177
Destructive Development Ignores Scientific Reasoning 183
Alternative Understanding of Economic Development 190
Tapping Creativity To Meet Real Needs 196
To Feed the Hungry and Protect the Environment 199
Forests, Wild Life and People 213
Future Development and Environment Protection 229
Gandhi Even More Relevant Today 260
People's Role Critical in Resolving Survival Issues 264

Acknowledgements *269*
Voices *271*
About the Author *273*

INTRODUCTION

In recent years it has become very clear that several environmental issues are increasingly taking the form of nothing less than a survival or an existential crisis.

On another front, not just nuclear weapons but also other weapons of mass destruction and the fast-emerging military use of robots/artificial intelligence also pose an existential threat. The environmental catastrophe resulting from actual use of nuclear weapons is one manifestation of how the existential threats on these two fronts can be interrelated.

The main focus of this book is on existential or near-existential threats and to emphasise the need for timely solutions before it is too late. These survival issues need to be resolved within a framework of democracy, peace and justice.

As the existing framework has failed to an alarming extent in resolving these issues, clearly big changes in global governance capable of resolving these issues are needed before it is too late. These changes will not come on their own; a huge and sustained upsurge in people's mobilisation on these issues all over the world is needed for this. On the one hand, this mobilisation will increase common people's consciousness about the urgency of these survival issues and on the other hand, this mobilisation will lead people to strive for a situation in which governments accord adequate urgency to survival issues.

An essential part of this mobilisation of people should be to bring justice and equality on the one hand and improve social relationships at all levels on the other hand.

Justice and equality will help all people to meet their basic needs without imposing a heavy burden on the environment. Improving social relationships will help to reduce significantly widespread distress and depression while leading towards highly creative relationships which can contribute so much to resolving the most critical issues in conditions of increasing happiness and enthusiasm.

In this mobilisation of people, the various movements for justice (including gender justice), protection of environment and all forms of life, peace and genuine spirituality need to come closer to each other. They have been fulfilling very important roles but also need to self-introspect, improve and innovate in close cooperation with each other. If this happens, then they can play a very crucial role in the overall efforts which are needed for resolving survival or existential issues in time.

All these efforts will keep alive hope in difficult times. The situation just now is very difficult but instead of saying it is too late for remedial action, we need to assert that it is never too late. This is the basic context in which this book is written.

ALARM BELLS RINGING

The World is not headed toward a sustainable future, but rather toward a variety of potential human and environmental disasters.

—The World Resource Report

The Earth is Finite
Survival Crisis like None Other

The authors of the 1992 declaration feared that humanity was pushing Earth's ecosystems beyond their capacities to support the web of life. They described how we are fast approaching many of the limits of what the planet can tolerate without substantial and irreversible harm.

... On the 25th anniversary of their call, we look back at their warning and evaluate the human response by exploring available time series data. Since 1992, with the exception of stabilising the stratospheric ozone layer, humanity has failed to make sufficient progress in generally solving these foreseen environmental challenges, and alarmingly, most of them are getting far worse.

—World Scientists' Warning to Humanity:
A Second Notice (2017), a Statement signed
by 13,524 signatories from 180 countries

> *If not checked, many of our current practices put at serious risk the future that we wish for human society and the plant and animal kingdoms, and may so alter the living world that it will be unable to sustain life in the manner that we know. Fundamental changes are urgent if we are to avoid the collision our present course will bring about.*
>
> —World Scientists' Warning to Humanity
> (1992 Statement of 1,575 top scientists)

THE MOST critical issue of our times is that the life nurturing conditions of our planet are being threatened by man-made factors as never before. The most discussed concern is climate change, but there are at least eight other serious environmental problems which can be treated as *survival threats*. In addition, there is the great risk from weapons of mass destruction including nuclear weapons, chemical weapons and biological weapons as well as the fast-emerging high threats from military use of robots.

These great risks increase further if we keep in mind the various interactions among them as well as their accumulated impact.

Such warnings have been voiced increasingly in recent decades by some of the most prominent scientists, and yet somehow these warnings have not led to the action that was needed to check these problems in time. Instead, despite the growing recognition of their seriousness, most of these problems have gone from bad to worse. In fact, during the last one or two years, the world appears to be moving away further than ever before from any credible response to these survival threats. And yet, we need to keep alive the hope for an adequate response to these threats before it is too late.

Warning to Humanity

In 1992, the Union of Concerned Scientists and 1,575 of the world's most prominent scientists (including 99 of the 196 living Nobel Laureates) signed a document called the *World Scientists' Warning to Humanity*. This document was sent to government leaders all over the world.

This document warned, 'Our massive tempering with the world's independent web of life—coupled with the environmental damage inflicted by deforestation, species loss and climate change—could trigger widespread adverse effects, including unpredictable collapses of critical biological systems whose interactions and dynamics we only imperfectly understand. Uncertainty over the extent of these effects cannot excuse complacency or delay in facing the threats'.

'The earth is finite', this document reiterates. 'Its ability to absorb wastes and destructive effluent is finite. Its ability to provide food and energy is finite. Its ability to provide for growing numbers of people is finite. And we are fast approaching many of the earth's limits. Current economic practices which damage the environment, in both developed and underdeveloped nations, cannot be continued without the risk that vital global systems will be damaged beyond repair'.

After listing the various serious threats, these 1,575 prominent scientists concluded, 'We the undersigned, senior members of the world's scientific community, hereby warn all humanity of what lies ahead. A great change in our stewardship of the earth and the life on it is required, if vast human misery is to be avoided and our global home on this planet is not to be irretrievably mutilated'.

Twenty-five years after this warning by leading scientists was voiced, another group of senior scientists supported by

13,524 signatories from 180 countries issued a follow-up statement titled: *World Scientists' Warning to Humanity: A Second Notice*. This follow-up statement reviewed the time series data on major problems and found that most of them are getting far worse. This statement said, 'Especially troubling is the current trajectory of potentially catastrophic climate change due to rising GHGs (Greenhouse Gases) from burning fossil fuels, deforestation and agricultural production—particularly from farming ruminants for meat consumption. Moreover, we have unleashed a mass extinction event, the sixth in roughly 540 million years, wherein many current life forms could be annihilated or at least committed to extinction by the end of this century'.

The statement concluded, 'To prevent widespread misery and catastrophic biodiversity loss, humanity must practice a more environmentally sustainable alternative to business as usual. This prescription was well articulated by the world's leading scientists 25 years ago, but in most respects, we have not heeded their warning. Soon it will be too late to shift course away from our failing trajectory, and time is running out. We must recognise, in our day-to-day lives and in our governing institutions, that Earth with all its life is our only home'.

Weapons of Mass Destruction

Nearly one hundred thousand people were killed within a few minutes in Hiroshima and Nagasaki after being hit by nuclear weapons in 1945. If we count the longer-term deaths caused by internal bleeding, leukemia and other forms of cancer, then the death toll is likely to be much higher. In addition, nearly seventy-five years later, the next generations continue to pay for this in the form of children born with mental retardation, physical deformities and other serious health problems.

Given this destruction caused by the first two nuclear weapons, it is truly scaring to know that today the world has around 15,000 nuclear weapons. And most of these nuclear weapons are much more destructive compared to the first two nuclear bombs. The US Department of State has provided the following data on the number of nuclear weapons with various countries: Russia-7,000, USA-6,800, France-300, China-270, Britain-215, Pakistan-140, India-130, Israel-80 and North Korea-10.

In addition, several countries are being mentioned as potential future processors of nuclear weapons; these include not just Iran but also possibly Saudi Arabia, Egypt, Japan and South Korea.

In a recent review of the nuclear weapon race, *The Economist* (27 January 2018) has noted, 'Both the USA and Russia are committed to costly long-term modernisation programs for their strategic nuclear forces that should ensure their viability for the rest of the century. Other states with nuclear weapons, such as China, Pakistan, India and, particularly North Korea are hard at work to improve both the quality and size of their nuclear forces'. Further, this review has noted that several disruptive new technologies and worsening relations among some of the nuclear weapon countries may also be increasing the risk of actual use of nuclear weapons.

In addition, despite the efforts to ban chemical and biological weapons, the threat of their highly destructive use still remains, not just by some countries but also by terrorists.

Some experts have warned that ultimately, military use of robots may prove to be the most destructive. Stephen Hawking, Elon Musk, Bill Gates and several others believe that, if adequate care is not exercised, general AI (Artificial Intelligence) can be an existential threat to the human race.

While all of these survival threats are important in themselves, their mutual interactions and accumulated impact can be much more destructive than the sum total of individual threats.

Challenge Ahead

The biggest challenge for human society in the next three decades or so is to significantly reduce all these survival threats, while at the same time meeting the basic needs of all people in a broad framework of justice, peace and environment protection.

Life without Domination and Discrimination

Development and Core Human Values

THERE ARE some core human values on the desirability and importance of which it is easy to get very widespread agreement in the entire world. These can be put together in numerous ways. Learned scholars will no doubt find very elegant ways, also very sophisticated ways of doing this. Here let us try to have a very brief and very simple formulation.

We should try always in our daily life and thinking not to intentionally cause any distress to any other living being. We should give a lot of effort and thought to this. There should be a firm belief in the well-being of all human beings (without any discrimination on any basis) and all forms of life, now and in the time to come.

Now this appears to be a very mundane statement, something we hear all the time. But look at the implications if most of the world's people actually practise this. Suddenly the world becomes a much better place, distress is greatly reduced and happiness increases in daily life. As the application of this very simple dictum becomes more and more thoughtful, other deeper changes also come at a wider level and for the longer term. The arms race reduces greatly, environment is much better protected, poverty decreases significantly, all sorts of crime and violence decrease even more significantly.

Unfortunately, what has been called 'development' did not give adequate attention to the protection and promotion of such core human values.

Instead, development was equated first and foremost with more and more of material goods. Then it was also linked with the enhancement of health and education facilities and services. However, if the core values have been neglected, then this by itself does not reduce distress at its deepest level. If all living beings and future generations are also considered, then development understood in this narrow sense, without the foundation of core human values, can also become the cause of increasing distress and very serious problems.

For the sake of understanding, let us consider two very different situations.

In one situation, there are the most modern and advanced goods and services and facilities, including artificial intelligence-aided robots, but the core human values indicated above are at minimal levels. In the second situation people live at primitive levels in forests, in the food-gathering stage, but there is a very strong presence of core human values among almost all the people. Of course, in such a situation there is complete illiteracy and there are no services and facilities but core human values exist almost everywhere.

What I want to emphasise is that if I was asked—which among the two situations I would choose (if such a choice had to be made), then without any hesitation and doubt I would choose the second situation as the better one. If the well-being and happiness of all forms of life and all future generations are considered, then this is better achieved in the second situation.

The discussion is presented in such a way to highlight and to make clear that in our bid to survive, the most important role is that of core human values. Human beings are very clever and capable. They are capable of advancing science and technology very rapidly and in amazing ways. But without certain core human values, the advancement of science and technology can easily turn destructive, disruptive and distressing. Hence, the most important base for humanity should be certain core human values which are in tune with the welfare of all forms of life and all generations. The wise scholars and sages of the world can be called upon to formulate these values in elegant and sophisticated ways, but people prefer these to be spread in more simple and interesting ways.

We Need this More than Ever Before

Urgency of Humankind's Protective Role

THE MILKY Way is only one of the countless galaxies which exist in the universe. The sun, our sun, is only one of the countless stars that exist in our Milky Way.

The Earth is only one of the planets (not to mention the several satellites, asteroids and comets) which form the family of the sun (the solar system).

And among all these, as far as present human knowledge goes, *life exists only on earth*. Of course, life may well be existing on several heavenly bodies, but as per the latest knowledge available at present to us, it is known to exist only on earth.

And what a diverse life it is. According to various estimates, there are over 13 million to 30 million species of various forms of life in the world. All these together contribute to making a beautiful world.

And among these, Homo sapiens or human beings are only one of the species, albeit the most capable one.

And what a complex structure each living being is. A newborn baby, for example, has over 2,000 million cells.

Among millions of life-forms, human beings alone have the capacity to work in a planned way for the welfare of all forms of life. Human beings alone have the capacity to work in a planned way for the protection of environment and habitats that sustain such diverse forms of life. Human beings alone can perceive the threats to the coming generations and take timely measures to protect future generations of human beings and other forms of life. And it is human being alone who have the capacity to seek destruction of this beautiful earth.

It is this unique capacity of human beings which defines their role on earth. The essential role of human beings on planet earth is to protect and promote the welfare of all life forms, including of course human beings, now and of future generations.

Whenever human beings have lived in greater conformity with this protective aim, there has been more happiness and less distress in this world. But when the tendency to deny and violate this aim dominates, the result is an increase in distress and destruction.

During the last century, the tendency to violate this aim has dominated. Further, massive technological changes have increased the capacity to cause distress and destruction to such an extent that for the first time in the history of the earth, human-made changes threaten the survival of many, perhaps most forms of life. Climate change and nuclear weapons (other WMDs or weapons of mass destruction) are just two manifestations of this destructive capacity.

This means that the need for establishing the protective role of humanity—so that human beings fulfill their essential role on earth—is greater than ever before.

For about 5 billion years the earth had existed in a lifeless state after its first independent existence from the sun. The first signs of life appeared millions of years ago in the form of slimy micro-organisms; this led to jelly followed by shellfish, vertebrate fish, and then reptiles. It is believed, the largest reptiles, the dinosaurs, were the dominant form of life on earth for 200 million years. Then suddenly, the dinosaurs disappeared—the reason is not firmly established yet, but it probably involved asteroids hitting the earth and huge volcanic eruptions.

Depending on who exactly we're willing to recognise as our first ancestor, human history goes back to anywhere between 40,000 years to 300,000 years or more. An element of DNA has been traced back to a common ancestor who lived (in Africa) about 300,000 years back. About 40,000 years ago the 'newest human model' (Homo sapiens) spread all over the world.

If earth's life is seen as a single day, then human beings proper appear only in the last second before midnight; yet their role on earth has been of a very critical nature as it is due to the activities of human beings that in 20th and 21st century AD, serious global threats to life-creating conditions on planet earth have appeared. Again, out of the nearly 40,000 years of Homo sapiens existence on earth, for nearly 39,800 years, humankind did not pose any such threat to the planet's life creating conditions. It is only during the last 200 years or so that Homo sapien's activities created such threats. Such threats increased the most in the 20th century, which is also the time when post-industrial revolution, humankind created weapons of mass destruction (WMDs), capable of destroying most forms of life many times over. Nuclear, chemical and biological weapons were actually used to kill several hundred thousand people. But it is ironic that humankind largely considers these last 200 years, and particularly the 20th century, to be the time when they have *progressed* the most or *developed* the most.

Today we are familiar also with the concept of *tipping point* according to which global warming beyond a point—a rise of over 2°C—may create conditions of a steady, unstoppable drift towards worsening disasters and catastrophic conditions.

Warnings about grave harm to life-creating conditions on earth have been voiced several times in the late 20th century, not to mention the first decade of the 21st century.

Despite such warnings, however, the predominant opinion among humankind is still to regard the last two hundred years (when the causes for such threats to life-creating conditions were created by humankind's activities) as the years of the *greatest progress* of humankind and indeed, as the period of greatest progress on earth. With such a worldview, isn't it most likely that what is considered as progress will be speeded up further, and in the process the life-nurturing conditions on earth will be further endangered?

This is a very important question for the future of all life-forms on earth and this is the reason why in any history of the earth, the history of humankind (despite its short duration in the earth's history of billions of years) deserves special attention and why the last two hundred years of humankind demand even more concentrated attention. Today, more than ever before we need to assert and establish the protective role of humankind on planet earth.

Threat to Life-Nurturing Conditions
Of Tipping Points and Planetary Boundaries

TODAY WE are familiar with the concept of tipping point according to which global warming beyond a point—a rise of over 2°C—may create conditions of an unstoppable drift towards worsening disasters and catastrophic conditions.

Warnings about threats to life-creating conditions on earth have been voiced in the late 20th century, and the first decade of the 21st century.

Edward Goldsmith and his co-authors warned in the widely discussed report 'Imperiled Planet' (MIT, USA): The danger is that we have gone beyond simply damaging ecosystems and we are now disrupting the very processes that keep the Earth a fit place for higher forms of life. For life, as we know it, to continue, the balance of gases in the atmosphere must remain within certain limits.... Beyond a certain point, the system may 'flip' to an entirely new state which could be extremely uncomfortable for life as we know it. At present, we have no idea whether we are heading for a 'flip' but, once triggered, the change to the new state could occur with extreme rapidity in perhaps as little as a few decades.

Harvard Professor Edward O Wilson, one of the world's leading experts on biodiversity, summarised the current state of other forms of life in an article in *Time* magazine. Biologists generally agree, he said, that, 'on the land at least and on a worldwide basis, species are vanishing 100 times faster than before the arrival of Homo sapiens'.

The United Nations Environment Programme issues periodic reports on the state of world's environment, recent trends and future prospects. The latest of these—Global Environment Outlook 5—has presented 'undeniable evidence that the world is speeding down an unsustainable path'. This report has voiced a clear warning that urgent changes are needed 'to avoid exceeding critical thresholds beyond which abrupt and generally irreversible changes to the life-support functions of the planet could occur'.

This issue of critical significance has been taken up in greater detail in the work of scientists at the Stockholm Resilience Centre (SRC). Johan Rockstrom, director of SRC says, 'The human pressure on the Earth System has reached a scale where abrupt global environmental change can no longer be excluded. To continue to live and operate safely, humanity has to stay away from critical "hard-wired" thresholds in the Earth's environment, and respect the nature of the planet's climatic, geophysical, atmospheric and ecological processes'.

The scientists at SRC first identified the Earth System processes and potential biophysical thresholds, which, if crossed, could generate unacceptable environmental change for humanity. They then proposed the boundaries that should be respected in order to reduce the risk of crossing these thresholds. The nine boundaries identified were: climate change, stratospheric ozone, land use change, freshwater use, biological diversity, ocean acidification, nitrogen and phosphorus inputs to the biosphere and oceans, aerosol loading and chemical pollution. The study suggests that 'three of these boundaries (climate change, biological diversity and nitrogen input to the biosphere) may already have been transgressed'. In addition, it emphasises that 'the boundaries are strongly connected—crossing one boundary may seriously threaten the ability to stay within safe levels of the others'.

To avoid transgressing planetary boundaries, while also meeting the basic needs of over 9 billion people in a sustainable manner during this century, we need to retain as much as possible the biosphere balance of the Holocene Approach of Earth History. The Holocene began almost 11,500 years back at the end of the last Ice Age. The gentle fluctuation in average global temperature and other favourable conditions all enabled humanity to develop agriculture and utilise earth's resources in other ways to meet the needs of a growing population. But humanity failed to realise the limits of this in good time, and even when it realised that it was overexploiting resources, humanity could not find the will and the capability to check this.

As a result, we have entered what many experts describe as the Anthropocene epoch, marked by the start of the nuclear testing in the 1940s. It's seen as the first epoch in Earth's 4-5 billion years during which *human beings or any form of life are the main causative agents of change in natural global system.*

The new epoch is full of big uncertainties, including survival of basic life-nurturing conditions. In the words of Rockstrom, 'The Holocene is the only equilibrium of the planet that we know for certain can support humanity as we know it. We have no evidence to suggest that we could morally and ethically support 9.5 billion citizens with a minimum standard of good life (outside of Holocene conditions)'.

The big question is—can humanity live up to the big challenge of protecting and maintaining basic life-nurturing conditions? This is a question which concerns all of us and we all need to get involved with this—for our own sake but even more so for the sake of our children and grandchildren.

What our Children will Inherit

A Destroyed Environment

WE MUST accord topmost priority to ensuring that we do not bequeath dangerous environmental conditions to the next generation.

Almost every one appears to be in agreement that providing a safe and secure future to our children (and grandchildren) is a matter of very high priority for them, perhaps the highest priority. Even people who may not have their own children are likely to have the same concern for the generations to come. Yet unintentionally and unwittingly, we become part of a systematic, relentless ecological destruction which, unless corrected in time, will almost certainly inflict grave harm on crucial life-nurturing conditions of the earth.

We often think of a secure future for our children in terms of some property and bank balance but these in themselves can never provide protection if clean and safe air, water and food are not ensured.

The United Nations Environment Programme issues periodic reports on the state of world's environment, recent trends and future prospects. The Global Environment Outlook 5 (GEO-5) report shows that already air pollution is among the main causes of premature deaths and health problems, especially in children. Poor water quality remains the largest cause of human health problems. The number of coastal dead zones has increased dramatically in recent years.

Summarising the results of its work, Stockholm Resilience Centre says, 'The group of scientists including Hans Joachim Schellnhuber, Will Steffen, Katherine Richardson, Jonathan Foley and Nobel Laureate Paul Crutzen, have attempted to quantify the safe biophysical boundaries outside which, they believe, the Earth System cannot function in a stable state, the state in which human civilisations have thrived…The scientists first identified the Earth System processes and potential biophysical thresholds, which, if crossed, could generate unacceptable environmental change for humanity. They then proposed the boundaries that should be respected in order to reduce the risk of crossing these thresholds'. In this context a key research paper titled, 'Planetary Boundaries: Exploring the Safe Operations Space for Humanity' authored by Johan Rockstrom and 26 other scientists has been published in the journal, *Ecology and Society*. This paper says, '…Anthropogenic pressures on the Earth System have reached a scale where abrupt global environmental change can no longer be excluded. We propose a new approach to global sustainability in which we define planetary boundaries within which we expect that humanity can operate safely. Transgressing one or more planetary boundaries may be deleterious or even catastrophic due to the risk of crossing thresholds that will trigger non-linear, abrupt environmental change within continental to planetary-scale systems...The exponential growth of human activities is raising concern that further pressure on the Earth System could destabilise critical biophysical systems and trigger abrupt or irreversible environmental changes that would be deleterious or even catastrophic for human well-being. This is a profound dilemma because the predominant paradigm of social and economic development remains largely oblivious to the risk of human-induced environmental disasters at continental to planetary scales...There is ample evidence from local to regional-scale ecosystems, such as lakes, forests, and coral reefs, that gradual changes in certain key control variables (eg biodiversity, harvesting, soil quality, freshwater flows and nutrient cycles) can trigger an abrupt system state change when critical thresholds have been crossed'.

What this paper says about air and chemical pollution has specific references to children. On air pollution it says that its various health effects '...convert to about 800,000 premature deaths and an annual loss of 6.4 million life years, predominantly in developing Asian countries. Mortality due to exposure to indoor smoke from solid fuels is about double that of urban air pollution (roughly 1.6 million deaths), and exposure to occupational airborne particulates accounts for roughly 300,000 deaths per year, mainly in developing countries'.

About chemical pollution, this widely acclaimed research paper says, '...A boundary focusing on effects of chemical pollution, on the other hand, could be based on reduced or failed reproduction, neurobehavioral deficits, or compromised immune systems, which are linked to the combined exposure to many chemicals. Such a planetary boundary would need to cover subtle effects on the most sensitive life stages in the most sensitive species and/or humans, with effects observable at the global scale. An example of this approach has been reviewed, based on the suggested increase in neurodevelopmental disorders such as autism and attention deficit and hyperactivity disorder (ADHD) in children. The widespread exposure to low concentrations of multiple chemicals with known or suspected neurotoxic effects may have created a silent pandemic of subtle neurodevelopmental disorders in children, possibly on a global scale. Of the 80,000 chemicals in commerce, 1,000 are known to be neurotoxic in experiments, 200 are known to be neurotoxic in humans, and five (methyl mercury, arsenic, lead, PCBs, toluene) are known to be toxic to human neurodevelopment'.

On a concluding note this paper says about the inadequacy of existing efforts to cope with new and extremely serious problems, 'Transgressing one boundary may, furthermore, seriously threaten the ability to stay within safe levels for other boundaries. This means that no boundary can be transgressed for long periods without jeopardising the safe operating space for humanity. Humanity thus needs to become an active steward of all planetary boundaries—the nine identified in this paper and others that may be identified in the future—in order to avoid risk of disastrous long-term social and environmental disruption…The knowledge gaps are disturbing. There is an urgent need to identify Earth System thresholds, to analyse risks and uncertainties, and, applying a precautionary principle, to identify planetary boundaries to avoid crossing such undesired thresholds. Current governance and management paradigms are often oblivious to or lack a mandate to act upon these planetary risks despite the evidence of an acceleration of anthropogenic pressures on the biophysical processes of the Earth System'.

Thus, from the point of view of our future generations, the most important issue is the many-sided ecological crisis which is now assuming the form of a survival crisis. If we are truly concerned about the future of our children (and their children) then we should start giving the topmost priority to these issues. When we compare this need with the petty issues which dominate the current discourse, we realise the urgency of genuine reforms. We owe it to our children and the next generation to initiate necessary changes and reforms before it is too late.

TIPPING POINTS

We stand on the brink of a global crisis in infectious diseases. No country is safe from them.

—World Health Report

Memories of Hiroshima

Risk of Nuclear Weapons Increasing

HIROSHIMA, 6 AUGUST 1945: Father Kleinsorge, a German missionary, heard pathetic voices of people asking for water. When he managed to reach the place from where the voices had come, he saw nearly twenty people, all of them in similar condition. Their faces were wholly burned, their eye sockets were hollow, the fluid from their melted eyes had run down their checks.

It is images such as these from Hiroshima and Nagasaki which have led people to conclude that the luckiest people in a place hit by an atomic bomb are those who die instantly.

Temperature at the hypocentre of the explosion reaching double of what it takes to melt iron, the face of a schoolgirl sitting almost a kilometer away from this hypocentre being burnt beyond recognition, skin sloughing off scalded bodies, badly injured starving people unable to swallow anything because of the stench of dead bodies—this was the devastation caused by a 12.5 Kiloton bomb in Hiroshima which killed and wounded as many people as a mass raid of 279 aircrafts, laden to capacity with bombs, striking at a city ten times as populous.

Nearly one hundred thousand people were killed within a few minutes in Hiroshima and Nagasaki after being hit by nuclear weapons in 1945, but if we count the longer-term deaths, the toll is likely to be as high as 350,000. So cruel was the devastation that all of us must necessarily ask ourselves—we certainly do not want Hiroshima to happen to our friends, but do we want it to happen even to our worst enemies?

Despite this, the incredibly cruel fact remains that humankind now possesses over 15,000 nuclear weapons which are many times more powerful than the ones used in Hiroshima and Nagasaki, and efforts are constantly on to increase the destructiveness of these weapons.

Even without the wartime use of nuclear weapons, merely maintaining a huge nuclear complex either for direct military purposes or for supporting it can pose a threat to the health and well-being of hundreds of thousands of people. As for the actual wartime potential of nuclear weapons, it is clear that a monstrous force several hundred times more destructive than what was seen in Hiroshima can be unleashed in a future nuclear war. The destructive potential of nuclear weapons is already adequate to destroy almost all life on earth, by its immediate effect and longer-term impacts of environmental ruin, cancers, genetic damage, starvation and worse. No matter where these are being produced—in which country and for what purpose, whether as deterrents or belligerents—there should always be only one answer to nuclear weapons—No, No, No!

What is more, the possibilities of actual use of nuclear weapons have been increasing in recent times. The concept of 'mutually ensured destruction' was supposed to be a big deterrent in preventing the actual use of nuclear weapons, but now other factors may overcome this deterrent role. As *The Economist* (27 January 2018) noted in its special report on future wars , the command-and-control systems of nuclear weapons are becoming vulnerable to hacking by new cyber-weapons or 'blinding' of the satellites they depend on. A country under such an attack could find itself under pressure to choose between losing control of its nuclear weapons or using them, the journal noted in its editorial.

This review noted that Russia and the USA are modernising their nuclear forces at huge expense and China is enlarging its nuclear arsenal. 'Existing nuclear arms control agreements are fraying. The protocols and understandings that helped avert Armageddon during the cold war have not been renewed'.

The Economist informed that the USA has been working on a concept known as Conventional Prompt Global Strike (CPGS) for over a decade. 'The idea is to deliver a conventional warhead with a very high degree of accuracy, at hypersonic speeds (at least five times faster than the speed of sound), through even the most densely defended space .... Russia and China claim that CPGS could be highly destablising if used in conjunction with advanced missile defences. Meanwhile, they are developing similar weapons of their own'.

The Economist review noted that both the USA and Russia are committed to 'costly long-term modernisation programmes' for their strategic nuclear forces that should ensure their viability for the rest of the century. Further, this review stated that other states with nuclear weapons such as China, Pakistan,

India, and particularly North Korea, are working hard to boost the quality as well as the size of their nuclear forces.

Tom Plant, a nuclear expert says, 'For both Russia and the US nukes have retained their primacy. You only have to look at how they are spending their money.'

James Miller, former under-secretary of defence for policy at the Pentagon, stated recently that the development of advanced cyber space, missile-defence, long-range conventional strike and autonomous systems, has the potential to threaten both sides' nuclear retaliatory strike capabilities, particularly their 'command and control', and that, 'the potential of a dispute leading to a crisis, of a crisis leading to a war, and a war escalating rapidly, is growing.'

Miller and Richard Fontane, the President of the Centre for a New American Security, have identified cyber and counter-space attacks, including use of satellite jammers, lasers and high-power microwave-gun systems, as a factor that can start an unplanned conflict.

James Acton, a nuclear policy expert at the Carnegie Endowment for International Peace, has listed some dangerous possibilities (i) command and control systems that are used in both conventional and nuclear attacks may result in confusing one with the other, (ii) an increased risk of cyber-attacks. If some counter-responses are delegated to autonomous systems capable of retaliating in just milliseconds, the risks increase further.

Another serious issue is that the overall possibilities of conflict involving nuclear weapon states, including possibility of conflict between great powers, has increased in recent times.

According to the 2017 year book of SIPRI (Stockholm International Peace Research Institute), at the start of 2017, nine states possessed 14,935 nuclear weapons, of which 4,150 were deployed with operational forces. Nearly 1,800 of these are kept in a state of high operational alert'.

This report says, both the USA and Russia have extensive and expensive programmes underway to replace and modernise their nuclear warheads, missile and aircraft delivery systems, and nuclear weapons production facilities. The nuclear arsenals of the other nuclear armed states are smaller, but all are either developing or deploying new weapon systems or have announced their intention to do so. China, India, North Korea continue to prioritise military nuclear programme as a central element of its national security strategy'.

According to this 2017 yearbook, 'Virtually all the major global indicators for peace and security have moved in a negative direction: more military spending, increased arms trading, more violent conflicts and the continuing forward march of military technology. Existing multilateral and bilateral arms control agreements and processes are also under challenge not least due to the deteriorating relationship between Russia and the US—raising questions of global concerns and potentially epochal scope'.

The Economist noted recently, 'The main reason why great power warfare has become somewhat more plausible than at any time since the height of the cold war is that both Russia and China are dissatisfied powers determined to change the terms of a Western devised, American-policed international order which they believe does not serve their legitimate interests'.

Things can get out of hand more easily in overall conditions of increasing tensions. As *The Economist* warns, 'The greatest danger lied in miscalculation through a failure to understand an adversary's intention leading to unplanned escalation that runs out of control. Competition in the "grey zone" between peace and war requires constant calibration that could all too easily be lost in the heat of moment'.

Beyond Science Fiction
Out of Control Robot Weapons

ONE OF the greatest upcoming risks of the next few decades is likely to come from robotic weapons, AI (artificial intelligence) weapons, or lethal autonomous weapons (LAWs). In August 2017, as many as 116 specialists from 26 countries, including some of the world's leading robotics and artificial intelligence pioneers, called on the United Nations to ban the development and use of *killer robots*. They even said that this arms race threatens to usher in the 'third revolution' in warfare after gunpowder and nuclear arms. They wrote, 'Once developed, lethal autonomous weapons will permit armed conflict to be fought at a scale greater than ever, and at time scales faster than humans can comprehend. These can be weapons of terror, weapons that despots and terrorists use against innocent population, and weapons hacked to behave in undesirable ways'.

'We do not have long to act', this letter warned. 'Once this Pandora's box is opened, it will be hard to close'.

Ryan Gariepy, the founder of Clearpath Robotics, has said, 'Unlike other potential manifestations of AI which still remain in the realm of science fiction, autonomous weapon systems are on the cusp of development right now and have a very real potential to cause significant harm to innocent people along with global instability'.

The Economist (27 January 2016) noted in its special report titled 'The Future of War', 'At least the world knows what it is like to live in the shadow of nuclear weapons. There are much bigger question marks over how the rapid advances in artificial intelligence (AI) and deep learning will affect the way wars are fought, and perhaps even the way people think of war. The big concern is that these technologies may create autonomous weapon systems that can make choices about killing humans independently of those who created or deployed them'.

This special report distinguished between three types of AI weapons or robot weapons (i) in the loop (with a human constantly monitoring the operation and remaining in charge of critical decisions, (ii) on the loop (with a human supervising machine that can intervene at any stage of the mission) or (iii) out of the loop (with the machine carrying out the mission without any human intervention, once launched).

Fully autonomous robot weapons (third category) are obviously the most dangerous. A letter warning against the coming race of these weapons was signed in 2015 by over 1,000 AI experts. An international campaign called 'Campaign to Stop Killer Robots' is working on a regular basis for this and related objectives. Elon Musk has pinpointed competition for AI superiority at national level as the 'most likely cause of World War III'.

A recent widely discussed review of new weapons said, 'The possibilities of killer robots can no longer be dismissed. Stephen Hawking, Elon Musk, Bill Gates and many other experts believe that, handled badly, general AI could be an existential threat to the human race', (a general AI machine would be able to carry out almost any intellectual task humans are capable of). Further this report notes, 'the biggest change in the way wars are fought will come from deploying lots of robots simultaneously' (*The Economist).*

Paul Scharre, an expert on autonomous weapons, has written that 'collectively, swarms of robotic systems have the potential for even more dramatic, disruptive change to military operations'. One possibility he mention is that tiny 3D-printed drones can be formed into smart clouds that can permeate a building or be air-dropped over a wide area to look for hidden enemy forces.

Several countries are surging ahead with rapid advances in robot weapons. In 2014, the Pentagon announced its 'Third Offset Strategy' with its special emphasis on robotics, autonomous systems and 'big data'. This is supposed to help the USA to maintain its military superiority.

In July 2017, China presented its 'Next-Generation Artificial-Intelligence Development Plan', which gives a crucial role to AI as the transformative technology in civil as well as military areas, with emphasis on 'military-civil fusion'.

If the arms race for AI weapons escalates, there will be a temptation all the time to actually use them to test their capabilities.

Peter Singer, an expert on future warfare at New America, a think tank, has said that very powerful forces propel the AI arm race—geopolitical compulsions, scientific advances and profit-seeking high technology companies.

Paul Scharre has also raised the possibility that perhaps because of badly written codes or perhaps because of cyber-attack by an adversary, military use of autonomous systems can malfunction, raising possibilities of attack on people or soldiers on the same side, or escalating conflicts or killings to unintended, highly exaggerated levels.

The Economist has written, 'The fast approaching revolution in military robotics is in a different league. It poses daunting ethical, legal, policy and practical problems, potentially creating dangers of an entirely new and, some think, existential kind'.

The UN's Convention on Certain Conventional Weapons (CCWs) has the mandate to prohibit or restrict some weapons which are recognised as causing unjustifiable suffering. Discussions about robot weapons and lethal autonomous weapons (LAWs) have been held at this forum, but clearly robot weapons cannot be regarded as conventional weapons.

A meeting of the CCW in November 2018 brought together government experts and NGOs.

The campaign called Stop Killer Robots wants a legally-binding international treaty banning LAWs. This is an issue which should get the increasing support of all people who believe firmly in peace and disarmament.

Answering the Tough Questions

Why Climate Change Needs Attention

IT IS widely and increasingly realised now that climate change caused by abnormally high concentrations of carbon dioxide, nitrous oxide, methane and other greenhouse gases (GHGs) can badly disrupt known life patterns of human beings and other forms of life within a few decades (certainly within this century). Already, this is seen to be an important factor in extreme heat conditions, other extreme weather events and disasters.

There is also fairly wide agreement now that the rise of global temperature related largely to climate change has to be somehow controlled at not beyond 2°C level.

It has been rightly pointed out, among others by representatives of island nations, that even a 2°C rise in temperature will cause fairly large-scale disruption in several life-sustaining activities apart from leading to the loss of vast low-lying areas (related to rise in sea-level) and worsening 'natural' disasters.

So ideally, we should have aimed at a target of 1.5°C instead of 2°C. But the sad reality is that even the target of 2°C is appearing very difficult today. So, reducing the limit to 1.5°C is not considered realistic.

It is important to at least achieve this 2°C target, because various projections show that beyond 2°C, the earth's natural processes begin to break down and cause more warming. Massive amounts of warming gases stored in the Siberian permafrost can melt and get released in the atmosphere. Humid rainforests can lose their humidity and begin to burn down, again releasing stored warming gases. Beyond such 'tipping points' the situation can get out of hand.

But at present are we on the path of achieving at least the 2°C target?

The Lancet Commission on Health and Climate Change (LCHCC), after examining a lot of latest data, has concluded in 2015, 'High-end emission projection scenarios show global average warming of 2.6°-4.8°C by the end of the century with all their regional amplification and attendant impacts'.

Further, this report (June 2015) with access to latest data says, 'GHG concentrations in the atmosphere are continuing to rise at a rate that is incompatible with limiting warming to 2°C in the coming 35 years (by 2050) and which exceeds the IPCC's "worst case scenario". We are on track for a global average temperature rise of more than 4°C above the pre-industrial temperatures in the next 85 years, at which point global temperatures will still be increasing by roughly 0.7°C per decade (due to the lag in reaching equilibrium). This distribution will not be even: the so-called polar amplification phenomena might cause temperatures in parts of the Arctic to increase by 11°C in this timeframe'.

The United Nations Secretary General's High-Level Panel on Global Sustainability (UNSHPGS) also made an equally disturbing observation in its report (2012): In its Emissions Gap Report of year 2010, the United Nations Environment Programme concluded that the currently forecast 2020 emission levels were consistent with pathways that would lead to a likely temperature increase of between 2.5° to 5° Celsius by the end of the 21st century.

The fact that the world's leading experts see current GHG emission levels and increases as going beyond the 'worst case scenarios' and certainly much beyond the 2°C limit is deeply disturbing. This implies that current life patterns are likely to be very badly disrupted, unless very significant and massive reductions can be achieved in the very near future, and then this trend can be maintained.

Counting carbon dioxide levels from 1870, the rise so far has been a little over 0.8°C. With global emissions increasing by about 50 billion tonnes a year, how many more years will it require to reach the 2°Celsius limit? 25 years? 20 years? What after that?

At this stage, the likelihood increases of non-linear, highly disruptive and catastrophic changes. As the LCHCC Report cited above says, 'As we proceed rapidly towards 4° Celsius warming by the end of the century, the likelihood of crossing thresholds and tipping points rises, threatening further warming and accelerated sea-level rise. Secondly, small risks can interact to produce larger-than-expected chances of catastrophic outcomes, especially if they are correlated. Such impacts (and their interactions) are unlikely to be trivial and could be sufficient to trigger a discontinuity in long-term progression of humanity'.

What is more, at this stage, the ability of the world community to take united action on an adequate scale itself will be greatly compromised. As this report continues, 'Indeed, on the basis of current emission trajectories, temperature rises in the next 85 years may be incompatible with an organised global community'.

Total carbon dioxide emissions since 1870 are estimated to be around 1,500 billion tonnes. *For 2° Celsius limit, these have to be kept below 2,900 billion tonnes. At present these emissions are increasing at about 50 billion tonnes in a year. Reducing this is very important, but what has been the recent record?*

The UNSHPGS Report quoted above says, 'Despite the adoption of the United Nations Framework Convention on Climate Change and its Kyoto Protocol, annual global carbon dioxide emissions from fuel combustion grew by about 38% between 1990 and 2009, with the rate of growth faster after 2000 than in the 1990s'. This shows a glaring failure of world leadership when faced with the most urgent life-saving tasks.

This report (2012) also said that the global carbon-dioxide reached 389 parts per million (ppm – million prats of atmospheric air) in 2010, and unless there is very significant shift in policy, is likely to exceed 450 parts per million over the coming decades. In 2015 the LCHCC reported, on the basis of more recent data, that carbon-dioxide concentration reached more than 400 parts per million in 2014. Every additional ppm is equivalent to about 7.5 billion tonnes of atmospheric carbon-dioxide. Carbon-dioxide remains in the atmosphere for a long time. 'Much of past emissions remain in the atmosphere and will drive continued warming in future'.

To what extent is climate change already with us? This report informs us that each of the last three decades was successively warmer than any preceding decade since 1850. It notes that 2014 was the hottest year on record. Arctic sea ice is disappearing at a rate of up to 50,000 square kms per year. The Antarctic ice sheet is now losing 159 billion tonnes of ice each year, and sea levels are rising inexorably. 'After only 0.8°C warming, many anticipated threats have already become real-world impacts'.

Finally, on the overall impact of climate change, the LCHCC report (2015) says that it supports the conclusion reached six years earlier by the UCL Lancet Commission in 2009—that climate change is 'the biggest global health threat in the 21st century'. On the broader impact of climate change this report concludes, 'No region is immune from the negative impacts of climate change, which will affect the natural world, economic activities and human health and well-being in every part of the world... The implications of climate change for a global population of 9 billion people threatens to undermine the last half century of gains in development and global health'.

Without Further Delay

As wildfires raged in the USA and Europe and around 125 deaths were reported from a heatwave in Japan, a study by leading experts from four institutions presented increasing evidence why the climate change and global warming prospects may be significantly worse than what has been commonly projected so far.

As we have said earlier, scientists from the Stockholm Resilience Center, the University of Copenhagen, the Australian National University and the Potsdam Institute for

Climate Research got together to prepare a report, published in the 'Proceedings of the National Academy of Sciences', which has raised the possibility of temperatures rising to 4 to 5 degrees Celsius even if current emission targets are met. This can happen if a critical threshold is crossed and several tipping points lead to abrupt change. This includes loss of Arctic summer ice, reduction of Antarctic sea ice and polar ice sheets, permanent frost thaw, loss of methane hydrates from ocean floor and weaker land and ocean sinks.

Johan Rockstrom, co-author of this report and director of Stockholm Resilience Center has said, 'These tipping elements can act like a row of dominoes. Once one is pushed over, it pushes earth towards another. It may be impossible to stop the whole role of dominoes from tumbling over. Places on earth will become uninhabitable if "hothouse earth" becomes the reality'.

He further said, 'We are the ones in control now. But once we get past two degrees, we see that the Earth tips from being a friend to a foe. We totally hand over our fate to an earth system that starts rolling out of equilibrium'.

Against these rising threats, what has been the level of response? The latest available information tells us that greenhouse gas emissions increased in 2017. In 2018, demand for coal increased after a gap of four years. Investment in oil and gas also increased at a world level while subsidies for renewable sources of energy showed a reducing trend in several places.

As *The Economist* pointed out, some of the obvious showpieces of success in confronting climate change have also turned out to be dubious ones. The million plus electric cars in China are known to obtain their energy from a power grid fuelled two-thirds by coal, so the net result may be more carbon emissions compared to fuel efficient petrol driven cars.

A recent study for South Asia has revealed that impacts of climate change are likely to be acutely felt over a much wider area than was commonly believed earlier. Muthukumara Mani is a lead economist of the World Bank and lead author of a recent World Bank publication, *South Asia's Hotspots: The Impact of Temperature and Precipitation on Living Standards*. He wrote recently, 'Today, almost half of India's population lives in areas that are projected to become moderate to severe hotspots by 2050, where the impact of climate change is likely to be the greatest. Although coastal areas receive much of the attention, we find that many of these climate hotspots turn out to be in inland areas—in central and northern India'. *Seven of the most affected districts are in Vidarbha region of Maharashtra State while three others are in Madhya Pradesh and Chhattisgarh. Other badly-affected areas are in Rajasthan and Uttar Pradesh.*

While climate change needs to get more importance in overall development planning, care should be taken to base this on justice and participation of people. About 50 years back, the top-down strategy of green revolution was introduced to overcome food shortages, supported by leading technocrats and international agencies, but this has turned out to be highly destructive for environment, health and small family farms. There is always a danger that any new threat, no matter how serious, can be hijacked by strong, influential vested interests to push their own selfish and dubious agenda. For example, GM (Genetically Modified) crops and related technologies have been pushed for their usefulness in times of climate change while all the evidence points that all the highly-damaging impacts of GM crops will be even worse in times of climate change as farmers need solutions that make them more self-reliant rather than more dependent on hazardous, expensive and ecologically disruptive technologies, controlled by a few big multinational companies.

Perhaps the only good news to emerge in recent times has been that public perception of climate change as a big threat is increasing. In 2017, a poll in 38 countries revealed that 61 per cent of people see climate change as a big threat. The next challenge is to take further this growing consciousness in terms of justice and equality-based solutions which can help to mobilise most people in the world on this and related survival issues.

Threat to All Forms of Life

Rapid Extinction of Species

A STUDY released in June 2015 has re-emphasised that the extinction of various species has escalated rapidly in recent times. In a widely discussed paper titled, 'Accelerated modern human-induced species losses: Entering the sixth mass extinction', Gerardo Ceballos, Paul R Ehrlich and others have stated, 'The oft-repeated claim that Earth's biota is entering a sixth "mass extinction" depends on clearly demonstrating that current extinction rates are far above the "background" rates prevailing between the five previous mass extinctions.

'Earlier estimates of extinction rates have been criticised for using assumptions that might overestimate the severity of the extinction crisis. We assess, using extremely conservative assumptions, whether human activities are causing a mass extinction. First, we use a recent estimate of a background rate of 2 mammal extinctions per 10,000 species per 100 years, which is twice as high as widely used previous estimates. We then compare this rate with the current rate of mammal and vertebrate extinctions. The latter is conservatively low because *listing a species as extinct requires meeting stringent criteria.* Even under our assumptions, which would tend to minimise evidence of an incipient mass extinction, the average rate of vertebrate species loss over the last century is up to 100 times higher than the background rate'.

Further, this study asserts, 'Under the background rate, the number of species that have gone extinct in the last century would have taken, depending on the vertebrate taxon, between 800 and 10,000 years to disappear. These estimates reveal an exceptionally rapid loss of biodiversity over the last few centuries, indicating that a sixth mass extinction is already under way. *Averting a dramatic decay of biodiversity and the subsequent loss of ecosystem services is still possible through intensified conservation efforts, but that window of opportunity is rapidly closing*'.

This study emphasises that there is no exaggeration, actually the reverse may be true. To quote:

> We emphasise that our calculations very likely underestimate the severity of the extinction crisis because our aim was to place a realistic 'lower bound' on humanity's impact on biodiversity. Therefore, although biologists cannot say precisely how many species there are, or exactly how many have gone extinct in any time interval, we can confidently conclude that modern extinction rates are exceptionally high, that they are increasing, and that they suggest a mass extinction under way—the sixth of its kind in Earth's 4.5 billion years of history...The evidence is incontrovertible that recent extinction rates are unprecedented in human history and highly unusual in Earth's history. Our analysis

> emphasises that our global society has started to destroy species of other organisms at an accelerating rate, initiating a mass extinction episode unparalleled for 65 million years. If the currently elevated extinction pace is allowed to continue, humans will soon (in as little as three human lifetimes) be deprived of many biodiversity benefits.

This view is supported by another widely discussed research paper titled, 'Planetary Boundries—Exploring the Safe Operating Space for Humanity' by Johan Rockstrom and 26 other scientists, who say: '...Currently, about 25% of species in well-studied taxonomic groups are threatened with extinction (ranging from 12% for birds to 52% for cycads). Until recently, most extinctions (since 1,500 AD) occurred on oceanic islands. In the last 20 years, however, about half of the recorded extinctions have occurred on continents, primarily due to land-use change, species introductions, and increasing climate change, indicating that biodiversity is now broadly at risk throughout the planet'.

This paper also indicates how biodiversity loss and extinction can lead to other catastrophic changes:

> Like land-system change, local and regional biodiversity changes can have pervasive effects on Earth System functioning and interact with several other planetary boundaries. For example, loss of biodiversity can increase the vulnerability of terrestrial and aquatic ecosystems to changes in climate and ocean acidity, thus reducing the safe boundary levels for these processes... Accelerated biodiversity loss during the Anthropocene is particularly serious, given growing evidence of the importance of biodiversity for sustaining ecosystem functioning and services and for preventing ecosystems from tipping into undesired states. A diversity of functional response

mechanisms to environmental variation among species in an ecosystem maintains resilience to disturbances. Consequently, ecosystems (both managed and unmanaged) with low levels of response diversity within functional groups are particularly vulnerable to disturbances (such as disease) and have a greater risk of undergoing catastrophic regime shifts.

The loss of species is also linked to other critical areas like climate change and global fresh water. Regarding the impact of climate change this paper says:

> Significant risks of deleterious climate impacts for society and the environment have to be faced even if the 2°C line (of temperature rise) can be held...Despite uncertainties related to the degree of hysteresis in the relationship between ice growth and ice creation in response to temperature change, raising CO_2 concentration above 350 ppm may lead to crossing a threshold that results in the eventual disappearance of some of the large polar ice sheets, with a higher risk of crossing the threshold as the CO_2 concentration approaches the upper end of the range...The contemporary climate is thus moving out of the envelope of Holocene variability, sharply increasing the risk of dangerous climate change. Observations of a climate transition include a rapid retreat of summer sea ice in the Arctic Ocean, retreat of mountain glaciers around the world, loss of mass from the Greenland and West Antarctic ice sheets, an increased rate of sea-level rise in the last 10–15 years, a 4° latitude pole-ward shift of subtropical regions, increased bleaching and mortality in coral reefs, a rise in the number of large floods, and the activation of slow feedback processes like the weakening of the oceanic carbon sink.

Clearly this has serious consequences for biodiversity and

species loss. As for changes in global fresh water use, this study says:

> The global freshwater cycle has entered the Anthropocene because humans are now the dominant driving force altering global-scale river flow and the spatial patterns and seasonal timing of vapour flows. An estimated 25% of the world's river basins run dry before reaching the oceans due to use of freshwater resources in the basins. Global manipulations of the freshwater cycle affect biodiversity, food, and health security and ecological functioning, such as provision of habitats for fish recruitment, carbon sequestration, and climate regulation, undermining the resilience of terrestrial and aquatic ecosystems.

As the most capable and powerful form of life on earth, human beings have a special protective role towards other forms of life, but instead they've adopted a highly destructive role.

We'll also do well to remember that all other forms of life, whether in the oceans or on land, whether wild or domesticated, have to bear the adverse impacts of the environmental ruin unleashed by Man; we human beings alone are responsible for unleashing this ruin. When our activities bring death and disease to them, we do not even hear their silent protests.

Slow but Certain Death
Risks of Hazardous Chemicals

THE NUMBER of chemicals in commercial use has been increasing at a very fast pace. Over 80,000 chemicals are registered for commercial use in the USA, and over 100,000 in the European Union. However, knowledge about the safety of these chemicals has lagged far behind the fast expansion of their use.

According to a widely discussed research paper titled, 'Developmental neurotoxicity of industrial chemicals' by Prof P Grandjean and Prof PJ Landrigan: Neurodevelopmental disorders such as autism, attention deficit disorder, mental retardation and cerebral palsy are common, costly, and can cause lifelong disability. Their causes are mostly unknown. A few industrial chemicals (eg lead, methylmercury, polychlorinated biphenyls [PCBs], arsenic, and toluene) are recognised causes of neurodevelopmental disorders and subclinical brain dysfunction. Exposure to these chemicals during early foetal development can cause brain injury at doses much lower than those affecting adult brain function. Recognition of these risks has led to evidence-based programmes of prevention, such as elimination of lead additives in petrol. Although these prevention campaigns are highly successful, most were initiated only after substantial delays. Another 200 chemicals are known to cause clinical neurotoxic effects in adults

This paper further says, 'The combined evidence suggests that neurodevelopmental disorders caused by industrial chemicals has created a silent pandemic in modern society. Although these chemicals might have caused impaired brain development in millions of children worldwide, the profound effects of such a pandemic are not apparent from available health statistics. Additionally, only a few chemical causes have been recognised, so the full effects of our industrial activities could be substantially greater than recognised at present. .....The few substances proved to be toxic to human neurodevelopment should, therefore, be viewed as the tip of a very large iceberg'.

The National Institute for Occupational Safety and Health in the USA estimated that some ten million workers in the USA are exposed to 11 high-volume carcinogens. Five to tenfold increases in cancer rates have been demonstrated in some occupations. According to estimates by the World Health Organisations, as many as 25 million agricultural workers in developing countries may be affected by occupational pesticide poisoning in a year. Polychlorinated biphenyls (PCBs), which were later linked to cancers and birth defects, had been earlier used not only in insecticide but even in kiss-proof lipstick.

A significant recent study has investigated the possibilities of cancer caused by commonly used chemicals. The subject of this study is obviously important as this has implications for millions of people across the world. This significance is also reflected in the fact that 174 scientists from 28 countries (including India) were involved in this study.

As many as 50 of the 85 chemicals studied were found to have a low dose effect on cancer-causing features. Thirteen had a threshold dose beyond which they started having cancer-generating effects. In this study, not just commonly used pesticides, insecticides, fungicides, but even additives to plastic, PVC and polycarbonate products (such as food containers and water bottles) and biocides used in hand washes and cosmetics were identified as having potential to cause cancer. Other commonly used items in daily life identified for such risks in this study include stain repellents used in fabrics and carpets, flame retardants used in paints, certain construction material and material used in aircraft as well as some commonly used drugs (such as phenobarbital and acetaminophen) and tricloscan (used in toothpaste).

The London Food Commission found that out of the 426 basic ingredients used in pesticides in the UK, 164 had been implicated in causing cancer, reproductive effects ranging from impotence to birth defects, genetic mutations or irritant reactions.

Keeping in view these high risks, more information about toxicity and hazards of various chemicals is badly needed. However, Russell Train, former chief of the Environmental Protection Agency, USA, has said, 'We have no way of systematically screening the chemicals that go into production; we have no way of knowing precisely which chemicals go into production every year. In other words, we not only don't know whether what's going on out there is dangerous—we don't even know what's going on out there. We have, however, learned one thing—it's what we don't know that can really hurt us, even kill us'.

Doctor Vyvyan Howard, former development toxico-pathologist at the University of Liverpool, UK, says in a paper published in *The Ecologist:*

> Recent research has shown that the synergistic effect among chemicals used in different combinations is much more dramatic than was previously thought. Yet we continue to test chemicals for their possible carcinogenic or mutagenic potential in isolation from each other. ....We have in out bodies today what is estimated to be between 300 and 500 chemicals that simply would not have been there 50 years ago, because at that time they did not exist or were present in the environment at undetectable levels. Trying to work out the toxicology of such a complex mixture presents many problems and renders the classical approach (of examining chemicals one at a time) quite useless.

According to the research paper by Grandjean and Landrigan quoted above:

> Testing of new chemicals before allowing them to be marketed is a highly efficient means to prevent toxicity, but has been required only in recent years. Of the thousands of chemicals used in commerce, fewer than half have been subjected to even token laboratory testing for toxicity testing. Nearly 3,000 of these substances are produced in quantities of almost 500,000 kg every year, but for nearly half these high-volume chemicals no basic toxicity data are publicly available, and 80% have no information about developmental or paediatric toxicity. Although new chemicals must be tested more thoroughly, access to these

> data can be restricted, because they could be claimed to constitute confidential business information. Absence of information about the neurotoxic potential of most industrial chemicals is therefore the main impediment to prevention of developmental disorders induced by neurotoxic pollutants. Accelerated testing of chemicals already in commerce is therefore essential. In the USA, a legal mandate to require testing was established in the Toxic Substances Control Act, but is largely unenforced.

Clearly while the health risks from chemical pollution are alarming, the lack of adequate information is also alarming. This knowledge gap must be filled up, while at the same time many-sided efforts to reduce chemical pollution should be made.

In addition, it should be emphasised that risks from ionising and non-ionising radiation can be very high and have been increasing. The threat of serious electro-magnetic pollution and related serious health hazards is very real, particularly for children. Irreversible genetic pollution from GM crops and serious health hazards of GM foods are also very worrying threats.

No Scientific Basis

GM Crops Extremely Harmful

GENETICALLY MODIFIED (GM) crops have been extremely controversial since their advent. Several reputed scientists have warned against their high risks and hazards. Ignoring the vast and serious evidence against them, powerful interests have been very active at high levels in India to get acceptance for fast spread of GM technology. In this context it is important to summarise the evidence against GM crops.

An eminent group of scientists from various countries who constitute the Independent Science Panel have said in their conclusion after examining all aspects of GM crops,'GM crops have failed to deliver the promised benefits and are posing escalating problems on the farm. Transgenic contamination is now widely acknowledged to be unavoidable, and hence **there can be no co-existence of GM and non-GM agriculture**. Most important of all, GM crops have not been proven safe. On the contrary, sufficient evidence has emerged to raise serious safety concerns that, if ignored, could result in irreversible damage to health and the environment.

GM Crops should be Firmly Rejected Now

In 2009-10, when the Bt brinjal debate had peaked in India, 17 distinguished scientists from Europe, USA, Canada and New Zealand wrote to the Prime Minister of India warning against 'the unique risks (of GM crops) to food security, farming systems and bio-safety impacts which are ultimately irreversible'. This letter adds, 'The GM transformation process is highly mutagenic, leading to disruptions to host plant genetic structure and function, which in turn leads to disturbances in the biochemistry of the plant. This can lead to novel toxin and allergen production as well as reduced/altered nutrition quality'.

> ...The basic problem is that GM as employed in agriculture is conceptually flawed, crude, imprecise and poorly-controlled technology, that is incapable of generating plants that contain the required multiple, coordinately regulated genes that work in an integrated way to respond to environmental challenges.
>
> ...GM has not increased yield potential. Yields from GM crops to date have been no better and in the case of GM soya have been consistently lower.
>
> ...GM crops have led to vast increases in pesticide use, not decreases and therefore reduction of agricultural pollution cannot be claimed.
>
> ...Climate change brings sudden, extreme, and unpredictable changes in weather, which requires that a

> cropping system be flexible, resilient and as genetically diverse as possible. GM technology offers just the opposite.
> ...Stability of productivity and production is much lower with many of the GM crops commercialised today. Herbicide tolerant GM soya is far more sensitive to heat or drought stress than conventional soya.
> ...GM crops are designed to be used in conjunction with synthetic pesticides and fertilisers, which are manufactured from oil and natural gas.
> GM crops do not reduce greenhouse gas emissions.
> Recent data from the US Department of Agriculture has shown a vast increase in herbicide use since the introduction of GM crops tolerant to the application of these agrochemicals.
> Therefore, the introduction of GM crops has exacerbated rather than reduced agriculture's carbon footprint and is clearly unsustainable.
> Alternative proven technologies that can reduce the amount of fossil fuel used in farming already exist. This includes methods for reducing fertiliser applications, selecting farm machinery appropriate for each task, managing soil for conservation, limiting irrigation and (using) agro-ecological farming techniques.

Several scientists involved in studying the implications and impacts of genetic engineering got together at the International Conference on 'Redefining of Life Sciences' organised at Penang, Malaysia, by the Third World Network. They issued a statement (the Penang Statement, 1994) which questioned the scientific basis of genetic engineering. This statement said:

> The new biotechnology based upon genetic engineering makes the assumption that each specific feature of an organism is encoded in one or a few specific, stable genes, so that the transfer of these genes results in the

transfer of a discrete feature. This extreme form of genetic reductionism has already been rejected by the majority of biologists and many other members of the intellectual community because it fails to take into account the complex interactions between genes and their cellular, extracellular and external environments that are involved in the development of all traits.

It has thus been impossible to predict the consequences of transferring a gene from one type of organism to another in a significant number of cases. The limited ability to transfer identifiable molecular characteristics between organisms through genetic engineering does not constitute the demonstration of any comprehensive or reliable system for predicting all the significant effects of transposing genes.

Some GEOs (Genetically Engineered Organisms) have been made with virus or transposon vectors that have been artificially enhanced to become less species-specific. Since viruses and transposons can cause or induce mutations, there is the concern that enhanced vectors could be carcinogenic to humans, domestic animals and wild animals.

Persons with allergies may have legitimate concerns that with genetic engineering, once-familiar foods may be made allergenic. Furthermore, they will not be able to protect themselves if the foods are not labeled to state that they have been produced from genetically engineered organisms. Allergenic effects could be carried with the transgene or be stimulated by imbalances in the chemistry of the host plant or organism.

Another problem is that field workers or neighbours may develop allergies to insecticidal transgenic crops. For example, a spider venom expressed in sugarcane might block a metabolic pathway only in insects and not in

> humans, but humans can nevertheless develop serious allergies to some venoms.
>
> With genetic engineering, familiar foods could become metabolically dangerous or even toxic. Even if the transgene itself is not dangerous or toxic, it could upset complex biochemical network and create new bioactive compounds or change the concentrations of those normally present. In addition, the properties in proteins may change in a new chemical environment because they may fold in new ways.

Many more opinions of environment scientists and groups of scientists can be quoted to make a strong case regarding the very serious hazards and risks of GM crops. As India's topmost expert on this subject, Dr Pushpa M Bhargava pointed out several years ago, 'Those who have favoured GM crops have been generally found to have close contacts with GM-promoting economic and corporate interests'. Hence it should be clear that the scientific evidence is clearly against GM corps.

The Growing Evidence
Hazards of Mobiles and Towers

MORE AND more evidence has been piling up about the health hazards associated with excessive and careless use of mobile phones as well as with the towers installed for mobile phone communication. The debate on the many-sided threats from electro-magnetic pollution is likely to hot up with the passage of time and we should not lose any further time in becoming more careful about avoiding, as much as possible, or minimising the threat of serious health hazards.

While hearing a case on this issue in India in 2017, a bench of Chief Justice TS Thakur, Justices R Banumathi and Justice UU Lalit asked the contending parties to file scientific and research data on this issue. The Court also took note of a recent book by Justice JL Gupta, a former chief justice of a high court, in which he had written that he had developed a cancerous tumour due to excessive use of mobile phone.

According to a study in 2016, in a year, forty to fifty thousand people are diagnosed to be affected by brain tumour in India out of which about 20 per cent are children. The study showed a significant increase in the number of children affected by brain tumour. The *Asian Age*, while reporting this study on 10 July 2017 said, 'Doctors said that this could be attributed to long-term mobile use'. This report quoted in particular, Dr PK Sethi, Professor and consultant of the neurology department at the Sir Ganga Ram Hospital as saying, 'There is a lot of literature that establishes a link between mobile radiation and brain tumour. Mobile phones emit radiation from their antennas and kids are at high risk as they possess soft tissues near the ear'.

The debate over electromagnetic pollution and more particularly, the hazards of cellphone radiation, has been escalating in many countries, but the overwhelming tendency has been to try to silence the critics by saying that the research conducted so far has not provided any serious reasons for concern.

Is this true? What really do the existing research results reveal?

The most comprehensive review of the existing research in this field has been authored by Dr Devra Davis in her recent book, *Disconnect—The Truth About Cell Phone Radiation.* Dr Davis is an award-winning scientist, writer and President of the Environmental Health Trust. After reviewing mounds of research papers, Davis has concluded: If you consider all of the studies that have been published, most of them have not followed people for a decade. But if you examine only those studies that have analysed people for a decade or longer you find one thing—every single one of them shows that long-term heavy use of cell phones has increased the risk of brain tumors.

Another important conclusion drawn by Davis (who as a fond grandmother is specially concerned about children) is that children face the most threat from the increasing hazard of electromagnetic pollution. She writes, '....Children's skulls and bone marrow are thinner and much more absorptive than those of adults—a fact that explains why children's heads can absorb double or more the radio frequency energy of adults' heads....All are agreed—children's brains and skulls absorb at least twice as much radio frequency radiation as those of adults. Bone marrow can take in ten times more radiation in children than in adults, according to reports from Swiss scientists in 2010'.

This view is supported by a lot of other evidence. In 2001, a Commission of the Royal College of Physicians, chaired by Sir William Stewart, issued this report:

> If there are currently unrecognised adverse health effects from the use of mobile phones, children may be more vulnerable because of their developing nervous system, the greater absorption of energy in the tissues of the head, and a longer lifetime of exposure. In line with our precautionary approach, at this time, we believe that the widespread use of mobile phones by children for nonessential calls should be discouraged. We also recommend that the mobile phone industry should refrain from promoting the use of mobile phones by children.

In this context, the evidence provided by a team of scientists based in Lund, Sweden, is also often cited. They note that, of course, the intense use of cell phones by children is unlikely to induce any obvious, dramatic, or immediate impacts on health. 'It may, however, in the long run, result in reduced brain reserve capacity that might be unveiled by other later neuronal disease or even the wear and tear of aging. We cannot exclude that after some decades of (often) daily use, a whole generation of users may suffer negative effects (when they are) maybe already in their middle age'.

One important view that has emerged in research by several scientists is that heavy exposure to electromagnetic pollution may result in breaking 'the blood-brain barrier', particularly in the case of children, exposing them to serious hazards. Devra Davis has carefully summarised these serious concerns, 'When it is healthy, the brain of rats and humans is full of fat and is well insulated to keep out anything that is watery. Careful analysis of the brains of the cell-phone exposed rats shows that after just a week of regular two-hour daily exposure, they release albumin—a critical material that should not be found escaping from the brain. As a result, these rats remain vulnerable to taking in other agents in the blood that normally would never enter their brain…Once the blood-brain barrier is breached, then anything circulating into our bodies at the time—alcohol, drugs, toxic chemicals, cigarette smoke, diesel exhaust, will more readily enter the brain from the blood'.

On the other hand, industry sources have steadfastly maintained that an overwhelming majority of studies conducted so far have not confirmed any serious hazards associated with the use of cell phones. It is true that many respected authorities have taken this view. That there are two sides to the controversy is accepted even by Dr Davis. As she puts it, 'Neither the danger nor the safety of cell phones is yet certain. How we manage that uncertainty could avert a global public health catastrophe'.

What she and some other researchers are saying is—It's true that hazards are not fully confirmed, but the available evidence is enough to call for many precautions.

This is also the view taken by two other authorities on this issue. Dr Elisabeth Cardis and Dr Siegal Sadetzki wrote in a publication of the British Medical Association, 'Simple and low-cost measures, such as the use of text messages, hands-free kits, and/or the loudspeaker mode of phone could substantially reduce exposure to the brain from mobile phones. Therefore, until definitive scientific answers are available, the adoption of such precautions, particularly among young people, is advisable'. This appears to be an eminently reasonable approach.

It is important to carefully examine all the existing evidence on this critical issue, particularly keeping in view the fact that children are being exposed all the time to more cell phone radiation and cell phone tower radiation not just for communication but also for entertainment and education. Children are exposed to very heavy radiation at a very tender age. In Delhi, cell phone towers are being installed even in parks where children come to play. Several researches have indicated that all this can lead to a public health catastrophe. The full implication of this may be revealed in about two decades or so. To prevent this catastrophe, we have to take preventive and precautionary action NOW—before it is too late.

Increasing Difficulties

Meeting Needs of the Grassroots People

IS IT possible for the world to meet the basic needs of growing population in the middle of fast increasing environmental stress? This is a crucial issue of our times and other critical issues like political instability and social unrest are closely linked to this.

The most likely estimates suggest that world population will grow from around 7.5 billion today to exceed 10 billion by the end of the century.

The United Nations Secretary General's High-Level Panel on Global Sustainability (UNSHPGS) wrote in 2012, 'The world's population now stands at over 7 billion. Another billion people are expected to arrive within the next 15 years or so.... The latest data suggest that the global population is likely to reach 9 billion people by 2040 and to exceed 10 billion in 2100'.

The struggle to meet the basic needs of all is likely to be most acute in least developed countries. These countries have a much higher population growth rate. These countries had a population of 832 million which, the panel said, is projected to grow to 1.26 billion people in 2030, an increase of 51 per cent in just 2 decades.

A majority of the world's people now live in cities. In 2010, the world's urban population was 3.5 billion. It is projected to go up to 4.9 billion in 2030.

The UN Panel pointed out that basic needs of significant sections of world population are still unmet.

- One billion people currently lack access to nutritious food.
- 2.6 billion are without access to basic sanitation.
- 884 million people lack access to clean water.

So, ensuring that basic needs of all people are met as the world population rises from 7.5 billion to around 10 billion people is going to be a huge challenge. This is made more difficult by two factors.

Firstly, in the past, several ecologically-destructive, non-sustainable paths have been pursued which are going to make future development achievements more difficult.

As the UN Panel stated, 'The loss of services derived from ecosystems is a significant barrier to the achievement of the Millennium Development Goals to reduce poverty, hunger and disease. Nearly two-thirds of the services provided by nature to humankind are found to be in decline worldwide. In effect, the benefits reaped from our engineering of the planet have been achieved by running down natural capital assets'.

Further, the Panel stated, 'If ocean acidification continues, disruption of food chains and direct and indirect impacts on numerous species are considered likely with consequent risk to food security, affecting the marine-based diets of billions of people seriously'. As climate change and related environmental problems are accentuated, livelihood and other systems critical to meeting basic needs of people can be severely disrupted.

What could be more easily managed was the reduction of inequalities and it is well known that reduction of inequalities can help greatly to meet basic needs of the poor. But even in this respect there have been glaring failures as huge inequalities have not just persisted but even widened.

The UN Panel noted several aspects of these inequalities.

- Gross national income per capita (based on purchasing power parity) in high income countries was about 5 times higher than in middle-income countries in 2010 and about 30 times higher than in low-income countries.
- Food wasted by consumers in high income countries (222 million tons) is roughly equal to the entire food production of sub-Saharan Africa (230 million tons).
- The irrigation required to produce the amount of food wasted annually is estimated to be equivalent to the domestic water needs of 9 billion people.
- Estimates suggest that as much as 80 million hectares of land (and the water that flows over or beneath it) have been acquired in new international investment deals since 2000, more than half of it in sub-Saharan Africa.

Oxfam has noted in its widely quoted report 'Even it up' (2017) that reduction of inequalities can contribute to meeting the basic needs of vulnerable people, but this potential has not been realised.

In 2014 a tax of 1.5 per cent on the wealth of billionaires could fill the annual gaps in funding needed to get every child into school and deliver health services in the poorest countries.

If India reduces inequality by 36 per cent, it can virtually eliminate extreme poverty.

In 2013, the world was losing $156 billion a year in tax revenue as a result of wealthy individuals hiding their assets in offshore tax havens.

In future projections of meeting basic needs of people, the difficulties and problems that are likely to arise in the future are frequently underestimated at present. Some environmental problems (not just climate change) are taking the form of survival issues, and unless checked soon, will make it more difficult for billions of human beings to meet their basic needs. Disasters linked to these threats will be a particularly serious problem. Hence, it is very important to make a realistic examination of the seriousness of the challenges ahead and strengthen the efforts for justice, peace and environmental protection which alone can provide a strong base for meeting the needs of all people.

Turmoil in Tender Age
Youth Depression

IT IS a telling comment on the contradictions of our development experience that in some of the most developed countries as well as developing countries recording the most high levels of economic growth, very high levels of depression have been recorded and that too particularly among the adolescents and youth. It is often said by adult citizens that most of their efforts of economic betterment are aimed at improving the prospects of their children, but if adolescents and youth are leading an increasingly depressed, insecure or uncertain life, then questions need to be raised about where actually society is headed.

According to a recent issue of *The Economist*, the number of American children and teenagers admitted to children's hospitals for reporting suicidal thoughts has more than doubled over the last decade. The suicide rate in the 15-19 age group in the USA increased by 31 per cent for boys and more than doubled for girls between 2007 and 2015. Suicide in the USA has become the second leading cause of death amongst college youth.

According to the national survey on drug use and health in the USA, in 2015, an estimated 3 million adolescents aged 12 to 17 (12.5 per cent of the age group) had at least one major depressive episode in the past year. California's biggest school district Los Angeles Unified tallied 255 incidents of suicide openness, self-harm and suicide attempts between 2010 and 2011, but in 2014 this jumped to 5,000. More than 30,000 students reported prolonged feelings of hopelessness and sadness; 9.1 per cent of middle students and 8.4 per cent of high students in the district actually attempted suicide.

The American College Health Association surveyed 100,000 college students at 53 US campuses and found that 84 per cent of US students feel unable to cope, 79 per cent are exhausted, 60 per cent feel very sad and more than half are experiencing overwhelming anxiety.

Another significant aspect of these trends is that girls are becoming much more vulnerable to depression compared to boys. According to the department of health and human services, the rates for depression among girls were more than double that of boys (19.5 per cent of girls, 5.8 per cent of boys).

In the context of UK, *The Independent* reported (27 February 2016) that rates of depression and anxiety among teenagers have increased by 70 per cent in the past 25 years. The number of children and young people turning up at hospitals with a psychiatric condition has more than doubled since 2009 and in the past three years, hospital admissions for teenagers with eating disorders have also almost doubled. In a 2016 study, 93 per cent of teachers reported seeing increased rates of mental illness among children and teenagers, and 90 per cent said the issues were getting more serious.

A study titled 'Mental Health among Youth in Sweden' by Mona Sommer (Nordic Center for Welfare and Social Issues) has compiled a lot of available data on this issue in the context of Sweden. The number of girls (15-19 years) who were treated in hospitals for depression increased eight times within 25 years. In a study of over a thousand Swedish young people in school years 7 and 8, 21 per cent of girls and 16 per cent of boys reported that they had harmed themselves more than five times in the past six months. Between 1991 and 2014, the sales of anti- depressants in Sweden increased by 13,690 per cent for girls aged 15 to 19, while that for boys increased by 6,710 per cent.

Very worrying depression rates are also being reported from several developing countries. On April 6, 2017, the WHO released a report titled, 'Mental Health Status of Adolescents in South East Asia—Evidence for Action'. This report covers 10 countries in the South East Asia region of the WHO, including India. Adolescents in this report are defined as children and youth included in the 10-19 age groups. These constitute 18.8 per cent of the population of these 10 countries, with a total number of 362 million. On the other hand, those in the 13-17 age group number 181 million, comprising about 9.4 per cent of the population.

In these 10 countries, in the age group 15 to 29, suicide is the second leading cause of death. The suicide rate per one lakh population in this age-group varies from 3.6 in Indonesia to 25.8 in Nepal to 35.5 in India (the rate in India is the highest amongst all the 10 countries). The percentage of adolescents who reported attempted suicides at least once in the past one year ranged from 3.9 per cent in Indonesia to 13.3 per cent in Thailand, while the average for all countries was 6.4 per cent. The percentage of students who reported feeling lonely most of the times or always in the past 12 months was 8.4 per cent for all 10 countries. Alcohol use at this tender age was as high as 23 per cent in Thailand and Bhutan. The country of Bhutan, famous for measuring Gross National Happiness, an index used to measure the collective happiness and well-being of its population, has a relatively high suicide rate at 11.7 per 100,000. Only 38 countries report having a national suicide prevention strategy.

In this study-overall depression rate was found to be very high for India. Even though the data for India in this report covers only a section of students for whom details were available, nevertheless it indicates disturbing trends. As many as 25 per cent of students were found to be suffering from depression as defined in the study. Eight (8) per cent suffered from anxiety and 10 per cent said that they had no close friends.

As evidence about the seriousness of the problem of youth depression has accumulated from all over the world, several studies have tried to identify causes and possible remedial actions. However, several of these studies tend to focus on only one or two of the more obvious aspects of a many-sided and deeply-rooted problem.

- A number of studies have highlighted aspects related to social media. While some studies mentioned overuse of social media, others quote stressed up teenagers as saying that it is a strain to be always available for social media. A very large number of adolescents complain about bullying on social media and receiving insulting as well as intimidating messages. In this context a study by Professor Jean Twenge of San Diego State University is often mentioned. Using data collected between 2010 and 2015 for more than 500,000 adolescents in the USA, this study found that adolescents who spent 3 hours or more a day on smart phones or other electronic devices were 34 per cent more likely to suffer at least one suicide-related outcome, including feeling hopeless or seriously considering suicide, compared to adolescents who use such devices 2 hours a day or less. Among adolescents who use electronic devices 5 or more hours a day, 48 per cent had at least one suicide-related outcome.

- The National Crime Records Bureau (NCRB) in its 2015 data made a shocking revelation that in India, one student commits suicide every hour. In India, suicide is the leading cause of death in the 15-39 year age group with 37 per cent of the total global suicide deaths among women coming from the country, according to a new study conducted by Indian Council of Medical Research (ICMR), Public Health Foundation of India (PHFI) and Institute for Health Metrics and Evaluation (IHME) in collaboration with the Ministry of Health and Family Welfare along with health experts and other stakeholders. About 800,000 people die by suicide worldwide every year—of these, 135,000 (17%) are residents of India.

Some other studies have time and again quoted teenagers who said that their school or college studies and extracurricular activities imposed a lot of burden and stress on them. Their perception of the exaggerated ambitions which their parents had from them, as well as their own excessive ambitions influenced by highly competitive environs, often raised education-related stress to levels with which they could not cope. Students took up courses on the basis of ambition rather than genuine interest and youths took up jobs also on this basis so that at a tender age, work became more of a stress than a joy.

Another factor cited is the significant erosion of the family support system which affects a large number of adolescents and youth today. In many families, both parents may be working and may have very busy work schedules as well. So, on the one hand several parents may not have the quality time badly needed by their children while on the other hand the inclination and the ability of several parents to handle the complex stresses of the young generation are also open to question. Hence the bond of trust which gives adolescents the confidence to share their life stresses does not develop along expected lines. Adolescents of present generation feel more lonely than any of their predecessors.

Early initiation of sexual activities before physical and emotional maturity has been reached has also been mentioned in some studies as one of the factors behind this problem. This has been found to have particularly strong impacts on girls.

These various factors have been emphasised in different studies which have often recommended remedial action based on their identification of a particular factor.

In addition, increasing the access to mental health services for adolescents is emphasised time and again not only in the context of developing countries but even in the context of those developed countries where the already recorded access is amongst the highest in the world.

While all these recommendations have their specific importance in certain contexts, these by themselves cannot provide the way out of the existing gloom for the youth and adolescents. Looking in isolation at one or two factors may lead to missing the wood for the trees. The real solutions can emerge only if the problem as well as its solution are related to a much wider context.

The wider social context has to be seen at two levels. On the one hand, as explained above, a lot of avoidable pressures have been built up very rapidly. On the other hand, society's base of an ethics-based value system has been allowed to erode very rapidly. It was this ethical value system which provided stability and guidance to choose between right and wrong, an ability which could be passed from parents to children and from teachers to students.

In the changed circumstances, many adults feel that they do not have the moral capacity to fulfill this role of a mentor and guide, while the youth suffer from the absence of value-based guidance at a time when they need this the most due to fast escalating pressure and uncertainties of their lives. The great vacuum created by erosion of ethical value systems sometimes gets hidden behind the glitter of many attractive-looking technological changes but the reality is that the collapse of value system creates such serious problems which no technological change can solve by itself.

So remedial action is also needed at two levels. Firstly, many needless and avoidable pressures which are creating problems for our youth should be re-examined so that what is not necessary and at the same time is harmful can be discarded. Secondly, what is most important is that the greatest attention should be given to creating a strong base of ethics-based value system in our society. Needless to add, this involves very basic changes in the present-day social system.

Sense of Urgency Missing
Failure in Critical Areas

AN EXTREMELY disturbing aspect seen at the world level in recent years has been that essential commitment to reduce the most critical issues could not be fulfilled. There are some crucial and critical issues like weapons of mass destruction, climate change and other environmental issues where planetary boundaries are being transgressed and tipping points are close to approaching. These are issues which can be called survival issues or existential issues. Unfortunately, even though the urgent need for timely and adequate action in these areas is well known, this was neglected, ignored, not prioritised or anyway the existing systems proved incapable of taking the necessary action which could prove effective and provide hope.

Even regarding the most critical areas of climate change and related issues, the Global Environment Outlook Report of the United Nations Environmental Programme has stated, 'A sense of urgency is lacking. Internationally and nationally, the funds and political will are insufficient to halt further global environment degradation and to address the most pressing environmental issues even though technology and knowledge are available to do so... The gap between what has been done this far and what is realistically needed is widening.'

In the context of the most critical problem of climate change, the United Nations Secretary-General's High-level Panel on Global Sustainability (UNSHPGS) has reported, 'Despite the adoption of the United Nations Framework Convention on Climate Change and its Kyoto Protocol, annual global carbon dioxide emissions from fuel combustion grew by about 38% between 1990 and 2009, with the rate of growth faster after 2000 than in the 1990s'.

The 2014 report of the IPCC has warned that global emissions since year 2000 have been rising ever faster at around 2% every year. In this dismal situation, the joint US-China agreement of 2014 to reduce emissions was presented as a ray of hope by some, but closer examination reveals that limits agreed to by these two biggest emitters were five to seven times the global average needed by mid-century, to be in line with the aim of limiting global warming to 2°C.

Besides, questions need to be raised whether such pacts between two major countries will harm the larger international process—which anyway is struck with many problems. As the Lancet Commission on Health and Climate Change (LCHCC) pointed out in their June 2015 report, 'the engagement of industrialised countries is weakening compared with the 1990s and early 2000s, with major emitters, such as Canada, Japan, Russia and, of course, the USA now operating only under the

Copenhagen Accord and Cancun Agreements whose targets are voluntary and not subject to common metrics'.

Within these countries (and also elsewhere, for example in some oil rich countries) there are big vested interests who have spent millions of dollars to hinder effective action on climate change. In 2010, when in the USA the House of Representatives was considering a proposal to cap US emissions, the lobbies of US industry spent close to $ 500 million on their (successful) campaign against this. *An important study has estimated that the Climate Change Counter Movement in the USA has an annual funding of around $ 900 million a year.*

The UNSHPGS also reported in the context of some other essential areas.

Around 13 million hectares of forest were converted to other uses or were lost each year from 2000 to 2010. Overfishing has led to 85 per cent of all fish stocks classified as overexploited, depleted, recovering or fully exploited, a situation substantially worse than two decades ago.

Meanwhile, agricultural run-offs mean that levels of nitrogen and phosphorus in the oceans have trebled since pre-industrial times, leading to massive increase in coastal 'dead zones'. The world's oceans are also becoming more acidic as a result of absorbing 26 per cent of the carbon dioxide emitted into the atmosphere, affecting both marine food chain and coral reef resilience.

On meeting basic needs of people, this High-Power Panel reported that the number of undernourished people in developing countries increased by about 20 million between 2000 and 2008. At the same time inequalities have increased. Average country-level income inequality increased around 20 per cent between 1990 and 2005.

A more detailed report of increasing inequalities was presented in Oxfam's report, 'Even it up—Time to End Extreme Inequality'. This report stated, 'Exreme economic inequality has exploded across the world in the last 30 years, making it one of the biggest economic, social and political challenges of our time. Age-old inequalities on the basis of gender, caste, race and religion—injustice in themselves—are exacerbated by the growing gap between the haves and the have-nots'.

This Report also pointed out,

- Seven out of 10 people live in countries where the gap between the rich and poor is greater than it was 30 years ago.
- In 2014 the richest 85 people in the planet owned as much as the poorest half of humanity.
- Since the financial crisis, the ranks of the world's billionaires have more than doubled; Forbes has pinned down a record 2,208 billionaires from 72 countries and territories.

Several aspects of accentuating inequalities have been becoming increasingly clear in recent years, but if any further proof of the worsening trends was needed, we can refer to the recent widely discussed working paper titled 'Indian Income Inequality 1922-24: From British Raj to Billionaire Raj', written by Thomas Piketty and Lucas Chancel.

This paper presents strong statistical evidence of the rapidly increasing income inequality in India. The share of the top 1 per cent part of the population in national income increased from 6.2 per cent in 1982-83 to 21.7 per cent in 2013-14 (the highest ever recorded share of the top 1 per cent in India). Income of the top 0.01 per cent of the country's population during the period 1980-2014 grew by 1,834 per cent, and the income of the top 0.001 per cent grew by 2,726 per cent. In sharp contrast, the income of the bottom 50 per cent of the population grew by only 89 per cent (containing all adults above 20 years of age) during this period.

During 1980-2014, the top 10 per cent of the population was able to capture about two-thirds of the total increase of income during this period, while the bottom 90 per cent of the population had to be content with only one-third of it.

These statistics for a leading country like India make it very clear the extent to which inequalities have escalated at a critical time when it was extremely important to reduce inequalities. What is more, this failure is linked closely to the larger failure of the economy to meet the basic needs of a very large number of people. This, moreover, is only one aspect of increasing economic inequalities. In rural areas, land redistribution has been almost entirely given up during the last 35 years; on the other hand, land has been taken from farmers for private industrialists, miners and real estate developers on a more massive scale than ever before. In urban areas increasing inequalities are even more visible than in rural areas. Huge loans taken by some of the richest persons or business concerns related to them have not been paid back to such a massive extent that it has created a big crisis—the crisis of NPAs (Non-performing assets)in the country's banking system.

These trends of increasing inequalities lead to denial of resources for the most important tasks of fighting poverty and hunger while also retarding broad-based, sustainable economic progress in various ways. In India as well as in other countries, the failure to reduce inequalities and create an equality based society and economy is likely to bring more distress in future when the ecological crisis and related problems will become more grim.

Our Endangered Planet
Through the Eyes of Aliens

IN THE recent hit film from India titled *PK*, one reason why the audience found the alien from another planet played by Aamir Khan so interesting was the questions he raised about several incongruities and absurdities of our everyday life. These are now so much interwoven into our daily life that we take these for granted, but stand aside to reflect for a while, and these are actually revealed to be not just absurd but often even harmful.

PK, the alien, however was only commenting on the day-to-day life he saw around him on earth. Supposing some aliens from outer space also come to study the most crucial political, economic and environmental issues of our planet, then they are likely to confront us with much more serious distortions regarding the way in which the present day world is structured—distortions which are extremely threatening and yet we have learnt to accept them as a given fact of our life.

The aliens are likely to express their greatest surprise at the weapons of mass destruction (particularly nuclear weapons) stored in several parts of the world. They are likely to say, 'We just can't understand why you are so keen to manufacture and store something which can destroy your entire planet'.

When the study team of aliens looks up all the facts on climate change, it is likely to comment, 'We just cannot comprehend how the leaders of the earth can endanger the life-systems of their planet to such an extent, and then behave as though they have no responsibility for this? Why are the steps needed to check climate change not being taken? Why have greenhouse gas emissions continued to increase after their dangers have become evident. We just cannot understand how such dangerous things are allowed to happen on your planet? Do you have no concern for the life of your children and grandchildren?'

These aliens are likely to be quite upset by all the air and water pollution, 'When your air pollution is so harmful, why don't you take timely steps to check this? What have you done to your rivers and oceans? Of course, you know how critical these life-systems are, yet you care so little to protect them! We just can't understand why you continue to spend billions on weapons and atom bombs yet do so little to save life giving systems of land, rivers and oceans!!'

The scholars from another planet are likely to shake their heads in disbelief and exclaim, 'We just cannot see any sense in this. At some places a few thousand people are called a nation. In other places over a billion people or ten thousand times that number are called a nation. What is your criteria for defining a nation? Is there any criteria at all? Yet you fight such bloody wars over a few kilometres of borders? Then at the United Nations you give one vote to a nation with a few thousand people, and also one vote to a country which may have ten thousand times more people. Also, at the United Nations you give veto rights to those countries which make the most destructive weapons and invade others with no concern for international law. How does your system work? Is there a system at all?'

The aliens are likely to become quite anguished when they see hungry people, people sweltering in heat waves or shivering in cold waves under the open sky—people dying due to denial of basic medicines and health care. 'Hey, what kind of people are you? You go on building atom bombs but cannot give food to the hungry. You are building such towering skyscrapers but don't provide shelters to millions of homeless. Half of your population is obese due to overeating, half is hungry and malnourished. You spend billions on liquor and smoking, but cannot provide milk to hungry children? What kind of a world have you created on this beautiful planet?'

Alien researchers are likely to remain dumb with shock for some time before they can say, 'Can't believe this, can't believe this at all. This data on rapes, abuse of children, domestic violence! What kind of world have you created? You call this civilisation?'

Their anger may shoot up further and they may say, 'You say you are the most evolved species on this earth (man), but what have you done to all other animals, birds, fish and insects? Your own books say that the extinction rate of species is now hundred times more than it was before you human beings appeared on earth. Do you think that the earth exists only for you? Why have you killed billions of other animals and birds in the most cruel way?'

Then the aliens may reflect for some time before saying,

'How can we expect you to be kind to other forms of life when you can be so cruel to fellow human beings? Where are the original people of America and Australia? Were not over 90% of them killed for greed and conquest? Why do you believe so much in conquest and domination? Why killing of other people is celebrated as bravery? Why do you always try to dominate others? Why do you discriminate so much among your own fellow human beings on the basis of colour, caste or religion? Why are you so often unjust to your own women folk?'

Then coming close to some concerns of PK, the aliens may say, 'You guys are divided so much on the basis of religion, while religion should show the path of working together for the welfare of others. This is spirituality—helping others with compassion—but in your case we daily see people being killed and injustice being done in the name of religion. How can this lead to spiritual progress?'

It is difficult to imagine what kind of reply the leaders of our world will provide to the study team from another planet, but one can imagine that it will be extremely difficult for them to provide any satisfactory answers. One can also imagine that the world's leaders will be most uncomfortable when faced with this barrage of comments and questions from the alien visitors.

But at least, we ordinary people should face this question honestly—isn't there a lot of truth in these comments and questions?

Even without any aliens questioning us, shouldn't we all worry about these basic issues? How can we go on squabbling over the most petty things while ignoring the basic issues? All this talk about PK-like aliens may be completely imaginary, but the issues and questions are real.

NEGLECTED PRIORITIES

When new laws and rules are enacted to address issues of violence against women or corruption, for some time there is a sense of achievement, particularly among those who have worked with good intention and dedication. But after a few years, honest introspection reveals that the problem has remained and may even have worsened as the forces that cause erosion of values are much more powerful.

Neither Worried Nor Bothered

Indifferent People

A VERY disturbing trend of recent times has been that while some very serious problems have appeared at the global level, somehow the overwhelming majority of the people of the world do not appear to be actively involved in any attempts to solve these problems in time. Indeed, most of them do not appear to be even very worried about these problems, despite the fact that they and their children are likely to be very adversely affected by these accumulating, accentuating problems.

Climate change and nuclear weapons are just two of these existential problems, but there are several others. The extremely serious nature of several of these existential problems (although not all) is widely known and recognised, yet somehow this has not led to the kind of mass public mobilisation on these issues that one would have reasonably expected.

Sustained mobilisation of people on such issues is badly needed for effective solutions to emerge. While this has not happened, short-term mass mobilisation of people, sometimes on relatively frivolous, contrived issues has been possible when this could secure behind-the- scene support of powerful forces. In some countries, this short-term mass mobilisation of people has proven powerful enough to even achieve the ouster of the entrenched political leadership and installation of new ones. In such movements, issues like fighting corruption and bringing a few democratic reforms have been loosely used, but after some time it is seen that things have not really improved even in terms of the narrow agenda of reform.

However, the real challenge anyway is a much bigger one—that of involving people sincerely in the most urgent priorities on a continuing basis. Without such mobilisation there is hardly any hope of resolving the most critical survival issues in time. Hence, it is very important to try to understand why such a mobilisation has not been possible so far and what can be done to make it possible.

When a large number of people are unable to meet their basic needs and life is a daily struggle to meet these needs, one can understand why these people are not inclined to give much attention to wider issues. This is particularly so when the movements to meet their needs do not make efforts to link this issue with wider existential issues.

But what about the majority of the world's people who are able to meet their basic needs and also have such access to educational opportunities and media as to become adequately aware of critical survival threats? Why do more of them not become adequately involved with resolving survival issues to any significant extent?

One important reason is the rampant consumerism and a widely pervasive culture of greed and instant gratification. On the one hand, this manifests in the craze of acquiring more and more material goods and on the other hand, in relentless pursuit of various sensual pleasures.

In such a situation there is hardly much inclination for devoting any time or thought to important survival issues. Another related important issue is the breakdown and turmoil in social relations which embroils people in entirely avoidable crisis situations in their daily life.

This takes up a lot of their time and also very adversely affects the potential of their creative strengths to accept important social responsibilities. These two related factors taken together create a situation in which more and more people get alienated from the most important issues instead of getting involved in resolving them.

Social relationships and social values are closely related to each other and these two, taken together in turn, have a huge influence on the ability of any society to engage in creative ways and on a sustained basis with its most critical problems. Thus, the inability of our society to engage creatively and effectively with our most urgent problems is not just a technological or a political failure. Rather, it is a much larger social failure, a failure of social relationships and value systems.

The failure to understand and acknowledge the much broader social dimensions of urgent survival issues has led to a situation where a holistic approach integrating solutions of these survival problems with many layers of broad-based social change could not evolve. This is the reason why people could not be mobilised on these urgent issues on a large scale in a sustained way.

At one time there were high hopes that as human society moves from a tradition bound patriarchal society to a more open society with many choices, the overall creative abilities of the society for resolving urgent problems will be automatically enhanced. It is true that the traditional society was encumbered by a lot of narrow-minded thinking and social relationships were strongly influenced by the urge to dominate others. The constraints of the patriarchal system made it difficult for the majority of members, particularly women and youth, to realise their aspirations. So significant changes in this tradition-bound society were certainly needed.

Unfortunately, however, the changes that have been unleashed in recent decades have been far removed from what was really needed. Strong trends of dominance still remain. There may certainly be more choices and more freedom but without improving the capacity for linking one's choices with what is good for society and without creating the overall social conditions in which better choices are likely to be exercised, the new changes can take society further away from resolving basic issues. The relentless consumerism and pursuit of sensual pleasures, aggravated further by aggressive commercialisation and increasing inequalities, have certainly increased the threat of social disruption and ecological ruin. In an unequal and unjust society, choices are being exercised in such ways that while the old stability no longer exists there are many-sided new complications in social relationships.

Hence, it is time to say that the pursuit of material consumption has to be disciplined by the needs of ecological protection and meeting basic needs of all. Similarly, the indiscriminate pursuit of sensual pleasures needs to be disciplined by considerations of ecological protection as well as stability of social systems. This is possible only on the basis of well-integrated suitable value systems which are emphasised at the family level as well as in educational institutions. Similarly, the seriousness of survival issues and other critical problems needs to be emphasised in educational institutions.

Such a deep concern for social harmony can help to create conducive conditions in which there is adequate space for creative engagement of people, particularly the youths with urgent survival issues on a continuing basis. The entire issue of resolving existential issues has to be well integrated with a wider process of social change which is conducive for people's close involvement with the most urgent issues. If adequate attention is given to such conducive social change, then the chances of people's actions to resolve survival issues in time will be greatly enhanced. A society which motivates and inspires people to take up the most urgent challenges in a spirit of cooperation is badly needed at this critical juncture of human history.

Just Laws are Not Enough
Need for an Ethical Value System

CORRUPTION AND sexual violence against women are regarded as different issues, which they obviously are. But two factors are common to these issues. First, both have received a lot of attention in recent decades; many solutions have been tried but things have become worse in respect of both.

The second common factor is that the root cause of both is the steep erosion of ethical values in society. Without addressing this, other solutions prove to be a false hope or at best provide only temporary relief. This, in fact, is true of most other problems.

When new laws and rules are enacted to address issues of violence against women or corruption, for some time there is a sense of achievement, particularly among those who have worked with good intention and dedication. But after a few years, honest introspection reveals that the problem has remained and may even have worsened as the forces that cause erosion of values are much more powerful. What is more, the erosion of values can even lead to misuse of well-intentioned laws in surprising ways and this results in new problems and distortions in society.

One example is the increase in sexual violence against women after and despite the enactment of some strong laws and rules accompanied by administrative actions in India. Clearly, some other forces have been more powerful. Another example is the extent to which the dowry system has worsened in India after the enactment of strong legislation to curb it. In recent years, even in remote villages, people, particularly women say that the dowry problem has worsened. They feel that the burden of a daughter's marriage is now greater than ever before in the face of increasing marriage expenses. At the same time, misuse of dowry law has been reported, resulting in new problems including the tragic and wrongful imprisonment of some men and even elderly women.

Clearly, the expected positives of legislation were outweighed by the much higher impact of escalating greed. In fact, the rapid growth of greed, fuelled by an overwhelming sense of consumerism and instant gratification has been a dominant aspect of changing values in recent times.

While this is manifested in the endless desire to acquire more and more consumer goods, money and high value assets, the latter is also seen as the means of accessing more and more sensual pleasures in more and more expensive forms and with a strong feeling of instant gratification. With this rapid increase in greed with all its associated attributes, the ethical value system based on honesty has steadily eroded.

Of course, the forces of greed and honesty have always existed in society—the question is which force becomes more powerful in a particular phase and due to what causes. What has happened in our country and a large part of the world is that the forces of greed have become much more powerful and this is being facilitated in several ways by an unjust and unequal economic system that actually encourages the attributes of excess ambition, cut-throat competition and pursuit of selfish ends without bothering about the fairness of the means. Hence, the ethical base of society in which honesty and simplicity were respected gets replaced by new role models of luxury, glamour and acquisitiveness.

In such a situation, when honesty and integrity are emphasised only in the narrow sense of improving governance and even narrower sense of increasing tax collections, this by itself cannot go very far as at the same time values opposed to honesty are being promoted by the unjust system which promotes consumerism and greed in various ways. The political system functions in tune with the economic system despite occasional calls for morality which sound increasingly hollow.

The drift away from honesty is also reflected in social relationships and these come under increasing stress. The durability of relationships based on trust is replaced by increasing uncertainty, distrust, suspicion, jealousy and revenge. Hence, the change is conducive for the spread of crimes ranging from corruption to robbery to various kinds of violent assaults. What ought to be understood is that in some contexts, spread of violence is inherent in unjust and unequal societies and is in fact encouraged by powerful forces. It is well-known what kind of powerful forces are involved in perpetuation of various rackets relating to liquor, trafficking and worse.

At another level, violence based on hatred and discrimination based on narrow considerations of caste, colour, race, religion etc. is also promoted in direct and indirect ways by powerful forces to prevent the kind of unity among people that can change unjust and unequal systems. If a firm belief in peace and non-violence are essential precepts of an ethical value system, respect for all human life is the basis for this. The threat of such a value-based system is met by promoting many kinds of discrimination for narrow ends.

When values like honesty, integrity, cooperation and trust are emphasised in daily discussions, most people are willing to express their commitment to these values. But unjust systems function in ways that create increasing pressures as well as temptations to make increasing compromises with the system. These difficulties need to be recognised by those who want to make sincere efforts to check the rot and try to make credible interventions for protecting the basics of an ethical value system. Without taking up this basic task, we cannot find sustainable and satisfactory solutions to major problems which are a cause of enormous distress as well as ecological ruin.

First, we need to create awareness of the great need for this basic task, then we have to plan its basics and take it forward at various levels including formal and non-formal educational institutions, media including social media and various movements including sincere spiritual movements as well as movements for social justice. In fact, convergence and cooperation of sincere spiritual and social justice movements will be extremely useful for strengthening the base of ethical value systems.

The essentials of an ethical value system include truth and transparency, honesty and integrity, cooperation and compassion, respect for all human beings and indeed all forms of life which manifests in deep commitment to peace and non-violence in both action and thought. These basics can be worked out in greater detail, but it is also important to relate these to practical aspects of everyday life. The wider social results may take some time to fructify, but if people feel relief in personal life and relations by following ethical value systems, it will bring more people to this fold.

Educational Reforms

Prioritise Neglected Objectives

EDUCATIONAL REFORMS have been very widely discussed, yet it appears that some crucial aspects have been missed. This is apparent from the growing neglect of the most important objectives of education, and also the growing gulf between the real needs and the actual trends of education.

Certainly, the most important objective of education is to prepare a strong foundation of the most important values of humanity to prepare children to be good and responsible citizens of tomorrow's world. These values include a deeply-rooted sense of equality of all human beings; denial of any discrimination based on religion, race, colour, ethnicity, gender etc. care and compassion towards all forms of life, a deep commitment to environment protection and peace, a strong sense of justice, honesty, truth and commitment to helping others.

If a firm foundation of even the first mentioned value can become deeply rooted, then this alone can help to end much conflict in the world. If all these related values can be imparted even to a reasonable extent, then nothing can stop the creation of a better world. Such is the important role of education. Here we mean education not only at the school and college but also at the family and community level.

While the great significance of this objective of education cannot obviously be denied, in reality it is neglected all the time.

The second most important objective of education is to provide conducive and happy conditions in which the creativity of children can flower properly and adequately. This objective is of course accepted by educational institutions and quite a few of them insist that this is actually happening in their schools. However, when we look at the extent to which education is dominated by severe competitiveness and the race to get as much marks as possible, then in such conditions the objective of providing happy conditions for creativity of children to flower properly cannot get priority much less find its true expression.

It is extremely important for true reform of education that the two most significant objectives of education should be reclaimed and accorded their important place in educational reforms.

THE WAY FORWARD

Earth is for all forms of life, not just for human beings, and as the most capable species clearly our relationships with other forms of life should be one of protective care. Thus the call to protect Mother Nature.

Critical Century
Finding Hope amidst Despair

DURING THE first two decades of the 21st century, it has become increasingly clear that this is going to be the most critical century of human history in the sense that survival issues or existential issues created by human activities are likely, sooner than later, to have an overwhelming impact. Among environmental issues this includes climate change or at least eight or nine other issues (like freshwater shortages, species extinction and drastic changes in oceans as well as land use).

Among security issues, this includes weapons of mass destruction (like nuclear, chemical and biological weapons) plus cyber warfare and robotic weapons (or AI warfare) which are dangerous enough in themselves but also enhance the threat of other WMDs.

Secondly, it is equally clear that so far, the preparations

to meet these extremely tough challenges of existential or survival threats have been alarmingly inadequate. Whether we see the progress in checking greenhouse gas emissions, or in climate change adoption measures, or in eliminating nuclear weapons, or in stopping military use of robots—all of these are most critical survival issues, yet the achievements in terms of resolving these issues in time have been shamefully inadequate.

This raises the question of whether remedial action is still possible or whether it is already too late.

We may already be late in some contexts, but surely in several other contexts it is still possible to do much. On the whole, rather than say that it is too late, it is better to say that it is never too late. While it is realistic to say that the overall situation is depressing and extremely difficult, we should nevertheless make full use of the openings that still exist. At no other time in human history was there greater need to find and utilise all openings of hope despite the widely prevalent despair.

More specifically, an extremely difficult but not impossible task is to resolve the survival issues in time within a framework of peace, justice and democracy.

To achieve this, the path ahead involves:

- A great upsurge among people of the world and a huge increase in their acceptance of this great responsibility.
- The coming together of peace, environment, justice, equality and gender movements to achieve this great task.
- The linking of local with global issues to obtain involvement of common people in these tasks.
- Linking up of the issues which help to create happier and more responsible life patterns, with the big survival responsibilities so that both can be achieved together.

- On the issue of resolving survival issues before it is too late, there needs to be growing unity of the world's people, cutting across all divisions and boundaries including national boundaries.
- Keeping in view the needs of resolving survival issues in time, mechanisms and world institutions need to be established to take the necessary, well-thought-out decisions in good time—the beginning of a democratic world government focused at least on a dozen or so critical survival issues that are carefully identified.

Change the Economic Model

A Distress Minimising Model of Development

THE OBSESSION with GNP-maximising model of development has led to the neglect of very important social, ecological and spiritual issues which can prove very costly and harmful. Hence there is a yearning among increasing number of people for alternative models of development and in this context, here we propose a distress minimising model of development.

The defining feature of this model is that here, the reduction of distress is the core issue; it is at the centre of all other discussions and questions. While in a GNP-maximising model the ways and means of maximising GNP growth constitute the core issue around which other issues revolve, in our model distress reduction is the central issue and other issues are examined only in relation to this central issue.

In this model, in the economic context, the main issue relates to the satisfaction of basic needs of all people. Whether very rich people get more income or very poor get more income may be the same in the model of maximising GNP, but this is not at all the case in the model of minimising distress. Increasing income of the poor contributes greatly to reducing distress but increasing income of very rich sections does not contribute at all to reducing distress. Hence increasing the income of the poorest sections becomes a crucial issue in the model of minimising distress. Similarly, the reduction of inequalities also becomes a crucial issue in this model.

At the same time, distress and happiness in daily life are very frequently related to our close social relationships—our families, close friends, workplace colleagues and neighbourhood relationships. How these relationships work out on a daily basis often has the most influence on happiness and/or distress that we experience in a day, and the happiness and/or distress that we contribute or cause to others in a day.

So, a key question is, in what ways our social relationships are likely to cause distress to us and to others and how to try to reduce this possibility as well as increase the happiness in social relationships. One example is that better understanding can lead to improved gender relationships, contributing enormously to reducing distress. Similarly, there are enormous possibilities of reducing distress in other relationships. Relationships of dominance should end, while relationships of care, compassion and cooperation should be encouraged in various ways. Fiercely competitive relationships need to be replaced by relationships of close cooperation.

Improving social relationships is one of the most important tasks, yet one of the most neglected. The effort should be to improve relationships at a deeper level and in durable ways, and not merely in terms of making a show of it. The improvement of social relationships should be an integral part of education both at the family and school (plus college) level.

Two important clarifications need to be made at this stage. In this model, do we include only reduction of distress of human beings, or are our concerns wide enough to include distress of all forms of life? Other forms of life—whether mammals, birds, amphibians, aquatic, insects, reptiles or others also feel pain and suffering. There should be no doubt that we should give adequate attention also to reducing their distress. Earth is for all forms of life, not just for human beings, and as the most capable species clearly our relationships with other forms of life should be one of protective care. Thus the call to protect Mother Nature.

The other classification relates to time-periods. Should our concerns be limited to minimising distress in present times or should we be equally concerned with reducing distress in future, or reducing the distress of future generations (ie. our children and grandchildren etc.)?

Clearly our concern should be to minimise the distress of future generations as well. This brings us very closely to the great need for concerns relating to sustainable development and environment protection which should be integral and very important aspects of the model of minimising present and future distress of all forms of life. Unfortunately, the very relevant word 'sustainable' is being reduced to a mere cliché but we use it here with all sincerity.

While we can try as much as possible to reduce distress, some forms of distress will always remain. We need to distinguish between avoidable and unavoidable distress. Almost all of us will experience the permanent loss of some near and dear ones, particularly those elder to us, in the course of our life. So, some distress is unavoidable. While trying to minimise distress, we also need the ability to bear unavoidable distress (or distress over which we have no control to reduce or limit it) with equanimity and without any breakdown, in the process also increasing the strength of others to do so.

This brings us close to wider spiritual concerns which need to be an integral part of any comprehensive effort to minimise distress. Spirituality is also of critical importance for understanding why and how any approach to life based on instant gratification and consumerism can actually result in increasing rather than reducing distress in various ways and at various levels.

Clearly the model of minimising distress provides a much more credible and reliable basis for human progress compared to the model based on maximising GNP.

Ensuring Disarmament
Plan to Reduce GHG Emissions

WHILE THE importance of reducing greenhouse gas emissions adequately and with the greatest urgency is widely recognised, this extremely important issue has not yet become an issue of mass mobilisation, particularly among weaker sections. The reason is obvious—people are too involved in their day-to-day problems. On the other hand, if any world-level planning for reducing GHG emissions is linked to meeting the basic needs of all people, then this will be a plan in which all weaker sections will have a vital stake. Hence, mass mobilisation on such a plan for meeting GHG emission reduction targets, while meeting basic needs of all, will be possible. This will also increase the opportunities for movements of justice and peace to work in close collaboration with the environmental movement.

Actually, as soon as such a plan is prepared, it becomes

obvious that in terms of the constraints of carbon space and other related constraints, there is no place in such a plan for extremely wasteful production and consumption as well as for weapons. Hence disarmament and concerns of the peace movement are built into such a plan.

The very pressing need to reduce greenhouse gas emissions adequately is now widely recognised, but a plan to achieve this needs to be integrated with a plan to meet the basic needs of all.

This plan should have the following components:

- All wasteful forms of consumption will be listed carefully and quantified. Those forms of wasteful consumption which involve particularly high GHG emissions will also be identified and quantified. A time-bound plan will be prepared to reduce all wasteful consumption as much as possible.

- In particular, an effort will be made to stop the production of all weapons to the maximum extent possible. In addition to the previously-known reasons for disarmament, we now have the additional reason of trying to curb all wasteful/harmful manufacture due to the urgent need to reduce GHG emissions.

- Possibilities of war and civil strife should be minimised, as apart from causing enormous distress to people, modern wars and the preparation for such wars involve a lot of GHG emission.

- A plan to increase the production of food, other goods and services to meet the basic needs of all in the world should be prepared, together with the most environment-friendly and least GHG emitting technologies that can be used for this purpose. This plan should be implemented in such a way as to ensure maximum local self-reliance in meeting basic needs, so that unnecessary transport is avoided, and maximum local employment is generated.

- The socio-economic changes that are needed to ensure that the adequate availability of all goods and services to meet all basic needs can actually be accessed by all should be identified and implemented.

- Energy planning should focus on replacing fossil fuels with solar, wind, hydrogen and other technologies which avoid or minimise GHG emissions and are environment friendly.

- All new technologies that are necessary for reducing GHG emissions and related objectives should be free from patents so that these can be used as widely as possible whenever needed. But technology transfer has its limits and local solutions for local problems should get maximum encouragement while keeping the door open for any input from outside when needed.

- As far as possible, no remaining natural forests should be cut. Timber needs should be curtailed as much as possible. Forest-dwellers or people living near forests should get first rights over minor forest produce, while also accepting responsibility for protecting forests. They should not be displaced but instead should be involved (with adequate incentives) in the protection of forests and wildlife.

- Very high priority should be given to reducing pollution and protecting habitats so that conditions for the healthy living of all life forms, whether on land or in water, can improve significantly.

- Farming and village-based life and livelihood patterns should get more help and priority compared to big industry and city-based life and livelihood patterns. Small farmer-based farming using environment friendly methods to produce healthy food with kind care of farm-animals should get the topmost priority and help.

- Top priority should be given to protecting fresh water sources and conserving water.

- All hazardous products, technologies, substances and chemicals should be carefully monitored and reduced.

An important role of this plan will be to make it very clear that if the constraints relating to carbon space (keeping emission levels low enough to restrict global warming at most to 2^{0}C) and other related constraints are to be respected, then there is no room for wasteful consumption and energy use, for weapons production, wars, deforestation and increasing exploitation of fossil fuels. Once the constraints are clearly understood, then the case for strictly using the available resources and carbon space for meeting only the basic needs of all becomes very compelling and strong.

Call for New Priorities and Solutions
Need for a World Government

IN THE middle of several problems which need attention, it is necessary not to lose track of what is most urgent. It should be possible to build a broad-based consensus that those factors which threaten the life-creating, life-nurturing conditions of our planet deserve the most urgent attention.

The most defining feature of the 21st century is the existence of such human-made threats to earth's life-nurturing conditions.

Such threats are related to climate change, crossing of other 'planetary boundaries' and 'tipping points' of life-giving conditions (including water and land use), accumulation of nuclear weapons and other weapons of mass destruction, as well as rapid spread of highly hazardous technologies and products particularly those capable of inflicting irreversible damage.

Edward Goldsmith and his co-authors warned in the widely-discussed report 'Imperiled Planet' (MIT, USA):

> The danger is that we have gone beyond simply damaging ecosystems and we are now disrupting the very processes that keep the Earth a fit place for higher forms of life. For life as we know it to continue, the balance of gases in the atmosphere must remain within certain limits. ...Beyond a certain point, the system may 'flip' to an entirely new state which could be extremely uncomfortable for life as we know it.

The World Health Organisation, in its annual World Health Reports (WHR), has warned time and again about the seriousness of global health challenges. The WHR has warned, 'We stand on the brink of a global crisis in infectious diseases. No country is safe from them.... Antibiotic resistance in hospitals worldwide threatens to leave medical and public health workers virtually helpless in the prevention or treatment of many infections'. Various factors have combined, this report says, to create, 'perhaps the richest opportunities ever for the spread of infections'. In particular, this report has focussed attention on the threat of new emerging diseases:

> During the past 20 years, at least 30 new diseases have emerged to threaten the health of hundreds of millions of people. For many of these diseases there is no treatment, cure or vaccine and the possibility of preventing or controlling them is limited....Any epidemic anywhere must now be seen as a threat to virtually all countries, especially those that serve as major hubs of international travel.

The Global Environment Outlook Report (GEO-1997) prepared by the United Nations Environment Programme has said on a note of warning,

> In the future, the continued degradation of natural resources, shortcomings in environmental responses, and renewable resource constraints may increasingly lead to food insecurity and conflict situations. Changes in global biogeochemical cycles and the complex interactions between environmental problems such as climate change, ozone depletion and acidification may have impacts that will confront local, regional and global communities with situations they are unprepared for. Previously unknown risks to human health are becoming evident from the cumulative and persistent effects of a whole range of chemicals particularly the persistent organic pollutants. The effects of climate variability and change are already increasing the incidents of familiar public health problems and leading to new ones, including a more extensive reach of vector borne diseases and a high incidence of heat related illness and mortality. If significant major reforms are not implemented quickly, the future may hold more surprises.

These warnings from authoritative sources clearly indicate the extent to which human beings, as well as all other forms of life are threatened, and the reasons why such threats to all forms of life and to life-creating conditions is likely to increase in future.

Faced with such an alarming situation, what has been the response of the world leadership so far? In the context of the most critical problem of climate change the UNSHPGS has reported, 'Despite the adoption of the United Nations Framework Convention on Climate Change and its Kyoto Protocol, annual global carbon dioxide emissions from fuel combustion grew by about 38% between 1990 and 2009, with the rate of growth faster after 2000 than in the 1990s'.

The GEO is therefore quite justified in lamenting that, 'A

sense of urgency is lacking. Internationally and nationally, the funds and political will are insufficient to halt further global environment degradation and to address the most pressing environmental issues even though technology and knowledge are available to do so... The gap between what has been done this far and what is realistically needed is widening'.

Greenpeace leader Jeremy Leggett has rightly said, '....the uniquely frustrating thing about global warming—to those many people who now see the dangers—is that the solutions are obvious. But there is no denying that enacting them will require paradigm shifts in human behaviour—particularly in the field of cooperation between nation states—which have literally no precedents in human history. This is the challenge. There is no single issue in contemporary human affairs that is of greater importance...'.

The efforts to reduce the risk from nuclear weapons have not made any significant progress. In fact, depleted uranium weapons are being freely used, and the chances of use of tactical nuclear weapons have increased. There is no guarantee yet that full-blown nuclear weapons will never be used. The nightmarish risks of terrorist groups gaining access to nuclear weapons have been discussed from time to time.

According to *Time* magazine, terrorist outfits, particularly when they have the help of a foreign government, can use nuclear weapons to kill around a hundred thousand people. The same magazine reported that the Aum Shinrikyo cult in Japan had accumulated material to manufacture chemical weapons which could kill 6 to 8 million people.

The US Air Force has estimated that in the event of a full-scale nuclear war between India and Pakistan, as many as 100 million people will die. The first hour of such a war itself would claim 17 million lives in Pakistan and 29 million in India. (Reported in *The Statesman*, 22 June, 1998).

Even without the wartime use of nuclear weapons, merely maintaining a huge nuclear complex either for direct military purposes or for supporting it can pose a threat to the health and well-being of hundreds of thousands of people.

These various threats to earth's life-giving conditions also cause massive distress to human beings as well as all other forms of life. The distress caused to other forms of life is not even mentioned even when countless birds fall dead in intensifying heat waves and prolonged droughts, or when millions of fish perish in rapidly drying and/or polluted water bodies. The fact that the extinction of species is now many times higher is absorbed just like that, without any intense feeling or sensitivity for the massive distress (and the associated injustice).

Both, due to the massive distress caused to various forms of life (including of course human beings) as well as (or even more) due to the often irreversible nature of the threat to life creating conditions on earth, the issues identified above should get the highest priority. While not many people question this, recent record confirms that effective and adequate remedial action is nevertheless very uncertain and in fact unlikely in the existing situation.

One problem is that decisions are taken at the level of various nation states, keeping in view the perceived self-interests of the country only. Secondly, the decisions within a country are often influenced too much by the powerful groups and strong lobbies keeping in view their own narrow interests. Quite often, some powerful corporate interests have a dominant say. At the international level, in the prevailing distribution of power, it is often what the few rich and powerful countries say which is more crucial for decision taking regardless of how unjust this may be. The more urgent concerns of the poorer and less influential countries are invariably neglected or accorded very low priority.

To what extent can the United Nations, as the most important international organisation of our planet, make up for these shortcomings in tackling the most urgent problems? The record so far is not at all encouraging, as is evident from the fact that remedial action on the most urgent issues has lagged far behind the real needs.

One serious problem of the United Nations is that in its General Assembly, a nation which has a population of one hundred thousand has the same representation (one vote) as a nation whose population is more than a thousand times.

The only privileged nations to have extraordinary (veto) power are the five permanent members of the Security Council who include the worst pollutants and human rights violators in the entire world. In several contexts, the form that the United Nations intervention takes in several crisis situations is closely linked to the foreign policy of the USA.

Unfortunately, narrow minded lobbies and vested interests have also become an integral part of the functioning of the UN, further reducing its ability to approach the most urgent issues in a non-partisan way which could get very widespread, broad-based support. Due to its increasingly bureaucratic ways, the UN has not been able to create and motivate a very wide and strong network of volunteers and activists to work on the most urgent issues.

The record of the United Nations can of course be improved for tackling the most urgent issues by initiating long overdue reforms. It can be repositioned as a more volunteer-based, activist-driven international organisation dedicated above all to the more urgent 'survival' issues. It can set a clear time frame for urgent actions on these 'survival' issues. But before this can get wider acceptance, the existing unjust structure of the United Nations (particularly the Security Council and the veto power for its five permanent members) must change. However, as things stand today, this appears to be very difficult.

So, the role of people's movements is a very critical one. Well-organized social movements with clear aims can mobilize people on a large scale to pressurise their national governments, the world leadership and the United Nations to initiate justice-based effective remedial actions on survival issues before it is too late. The movements for justice and equality as well as movements for peace and environment protection supported by movements for gender justice, animal rights movements and several other movements, all can contribute to this by better cooperation and integration with each other for wider objectives.

If such movements can create strong public opinion for justice-based solutions to survival issues, then true and sincere internationalism, with the objective of saving our earth and uniting all people for this can progress much beyond limited reforms of the United Nations. It should be possible then to think of creating a limited form of world government with a mandate for initiating adequate and effective remedial actions on the most urgent survival issues. In other words, certain critical existential and survival issues will be identified as the decision area of the world government so that critical decisions can be taken within time limits in the best interests of all life forms of our planet without being hindered by narrow national, selfish and power considerations. The mandate is to take these decisions within certain time limits and within the framework of justice, peace and democracy.

To start with, the mandate of the world government can be restricted to urgent survival issues like climate change, weapons of mass destruction and a few other such issues. To form the world government, one representative can be selected from a constituency of about 20 million people, ignoring national boundaries. The election campaign should be conducted mainly on urgent survival issues. In addition, various professional bodies (such as organisations of climate scientists, nuclear scientists, medical scientists etc.) should be asked to select their most capable representatives to assist the world government.

The tasks of the world government can be the following:

- Ensuring complete prohibition of nuclear weapons or any other WMD or existential threat weapon in the arsenal of any country. However, the world government will be entitled to stock limited, say about 2% of the existing stocks, at 3 or 4 places in the world to check any efforts of terrorists to acquire and misuse such weapons.

- A detailed justice based agenda for checking climate change and all other existential environmental problems to tolerable limits should be prepared and enforced by the world government. Strong time-bound action should be taken on other environmental issues which pose survival threat. The industrial countries with their historical responsibilities should be asked to accept justice based financial responsibility for reducing GHG emissions.

- Over a period of time, all national armies and weapons expenditure should be curtailed by about 90 per cent or so. The task of border disputes settlement should be taken up by the World Government. (Global production of all arms can be reduced by over 80 per cent or so.)

At the next stage, if the internationalism and one earth consciousness of people can continue to rise, the mandate of the World Government can be expanded to include additional issues. Gradually, our world should move from artificially created boundaries to one world, based on equality of all human beings at all levels and concern for all other forms of life. At the same time, however, decentralization and local self-government should be strengthened everywhere.

SOCIAL MOVEMENTS

The gulf war's oil fires spew noxious gases, soot and oil, mercilessly degrading the frail global eco-system. The burning wells and oil lakes released 50,000 tonnes of sulphur dioxide daily—a prime cause of acid rain—and 100,000 tonnes of sooty smoke. Meteorological satellites have detected smoke at a height of 6-7 kms and as far away as 2,000 kms from the source. Sunlight-absorbing plumes of smoke have been measured over an area of 42,000 square kms. The most dangerous aftermath relates to the release of greenhouse gases and sub-blocking soot production. The damage caused by the unhealthy smog and acid rain to trees, lakes, streams and rivers is incalculable and will continue long into the future.

—*Bernard Lowen*

Survival Issues, People's Response
Movements for Justice and Equity

SEVERAL SINCERE people's movements in various parts of the world have been playing an important role in reducing important problems and raising neglected issues. Some of the most important movements are movements for justice and equality at various levels (including gender), environment protection movements and peace movements.

While their role has been an important one, due to various reasons a decisive contribution has not been possible so far to the most critical issues which can be called survival issues or existential issues. Some of these issues include an almost complete ban on all weapons of mass destruction including nuclear weapons, checking military use of robots and significantly reducing the threat of climate change and several other related environmental problems before it is too late.

In terms of their capacity to cause the most destruction and distress these survival issues are becoming extremely important and people's movements have to become stronger, unite and bring other changes in time in order to be able to contribute significantly to resolving these survival issues before it is too late.

There is need for convergence and unity of various movements so that they can together contribute to resolving these survival issues. A very big challenge is to take these issues among common people in such ways that these can be mass mobilisation for resolving these issues within a framework of justice, peace and democracy.

One reason why adequate mobilisation could not take place on the most urgent survival issues is that the solutions could not be linked to needs and aspirations of people, particularly the weaker sections. In this context this writer had earlier proposed a plan which links the reduction of greenhouse gas emissions to the fulfillment of basic needs of all people and the reduction of weapons. Other proposals which have been made include the imposition of heavy taxes or fees on fossil fuels and then using the funds so obtained to meet the basic needs of weaker sections in eco-friendly ways.

No doubt many other creative ways of linking solutions of the most serious environmental problems with reducing poverty and deprivation can be made. Similarly, the savings linked with the phasing out of weapons of mass destruction can also be used for meeting the basic needs of people. In many creative ways the actions against survival threats can be given such a form that these also help to reduce the immediate distress and deprivation of people. Of course, a lot of careful planning is needed for this, but this is a real possibility. The reduction of fossil fuels has to be handled carefully to avoid temporary disruptions and for this, firstly fast and adequate growth of eco-friendly alternatives is necessary. This is difficult but achievable.

When solution of survival issues is linked to immediate reduction of deprivation then more people can be mobilized for actions relating to such solutions. In the course of this mobilisation they also become more aware of survival issues. Hence, more and more people become committed to resolving survival issues in the near future. With the dedicated involvement of increasing number of people, more creative solutions start emerging at the grassroots level. A situation is created of creative and enthusiastic involvement of people in finding solutions for the most urgent problems, replacing the earlier apathy and alienation.

What is more, there are other possibilities of making this creative process a very happy one. Quite often, the task of combating and checking very serious problems like climate change has been presented as a very burdensome responsibility for which people may have to accept several restrictions and sacrifices. This is why, in common imagination, this entire work is seen as one involving heavy responsibility and so there may be a tendency to shirk away from it. However, one can and one should take an entirely different view of the existing challenges to create a situation in which overall happiness can increase when people increasingly and willingly take up more and more responsibilities.

If we seriously examine the causes of human distress, then apart from socio-economic inequality and injustice, these are most often rooted in failures of human relationships at various levels. Most human relations are based on notions of dominance and self-interest while these need to be based on genuine sharing and cooperation. If along with linking the resolution of survival issues with reduction of poverty and deprivation, we can link this with well-thought-out efforts for basic improvements in human relationships at several levels, then just the right conditions can emerge for the most creative flowering of human talents for resolving the most urgent and complex problems in the middle of increasing happiness and enthusiasm of people.

It has often been observed that problems of human relationships are more likely to increase in a situation of absence of a larger purpose in life. On the other hand, when clear possibilities exist of making an important social contribution then possibilities of not only working with more commitment but also working in closer cooperation with others increase. This can be even more clearly observed among youth and in fact is very clearly visible among children. So, if such conditions can be created by appropriate social change and if, in addition, several meaningless restrictions placed on women and youth are removed, clearly the possibilities of human relationships to become creative and happy can increase significantly.

So instead of feeling too burdened by the enormity of the tasks ahead can we see a time when people will be very happily involved in sorting out these admittedly complex problems with a song on their lips and with music in their hearts?

Impact of Social Movements
Need for Better Linkages and Understanding

DESPITE THE increasing presence and recognition of social movements in all parts of the world, they have not yet been able to make any decisive impact on the resolution of most urgent issues. While respecting their many-sided creative and constructive contributions, at the same time the need for improvements and greater effectiveness should be re-emphasised, so that these social movements can actually contribute to resolving urgent issues like climate change and disarmament before it is too late.

These movements lack strong linkages with each other, often functioning in isolation. This can lead to a tendency to take more extreme views without recognising the need to adjust with other's highly desirable objectives. Also this means these movements cannot combine their strengths to have a higher impact on policy making.

This is unfortunate because, in reality, the most important movements for socio-economic justice or equality, environment protection and peace are closely related to each other. All the three movements need similar changes in value systems for their success—a change away from dominance and greed towards sharing and cooperation.

The perpetuation of economic inequalities is a constant source of tensions which can any time lead to conflicts. Whenever a large number of people are deprived of basic needs, they can easily be led towards violence. So reduction of economic inequalities and fulfillment of basic needs of all people will certainly lead to more conducive conditions for peace.

On the other hand, establishment of peace will certainly lead to more favourable conditions for meeting the basic needs of all. The annual military expenditure of the world is estimated to be many times the additional annual expenses of meeting some of the most pressing needs of poorer countries. Hence a significant reduction in military expenditure can provide most of the resources needed to meet the basic needs of all people.

The Human Development Report (HDR) reported, 'The Gulf war is estimated to have cost $1billion a day. The cost of just ten days of war was enough to immunise all the children of the Third World against vaccine-preventable diseases over the next ten years'. If we count the long-term costs of treating injuries and illnesses caused by war, then the costs increase further. This is particularly true when weapons causing long-term damage are used, as in the well-established case of the massive use of depleted uranium in Iraq and deadly chemical weapons in Vietnam. The actual costs of the military build-up are pushed up due to yet another factor—the bias towards military research can distort the entire pattern of scientific research.

Heavy military spending has acquired a momentum of its own, and shows no signs of diminishing even if security threats diminish. According to the HDR, 'The huge commercial and military organisations are fairly inflexible and cutting arms production and procurement can bring serious problems'. The arms industry is known to pay scant regard to welfare aspects of its functioning and is full of murky, clandestine deals which ultimately lead to the avoidable loss of thousands of lives, mostly in developing countries. The HDR adds, 'Arms dealers continue to ship weapons to potential trouble spots, showing little concern about fanning the flames of conflict. Much of the pressure for international sales comes from producers promoting overseas sales to recoup overheads and maximise profits...Despite rhetoric to the contrary, the heads of some industrial countries take a keen interest in promoting international arms sales'.

Similarly, there are close linkages among the environment protection movement. Environmental destruction can increasingly lead to or contribute to conflict situations. Ecological ruin destroys livelihoods and this in turn leads to conflict. Even more significantly, it deprives people of the most basic of all needs—water, leading to water-conflicts among regions or nations. A senior UN official has raised the possibility of major wars being fought in future over the distribution of water.

On the other hand, war and conflict—as also the build up of arsenals of destructive weapons for war—lead to large scale environmental destruction. Matthias Finger writes in an article, 'The Military, the Nation State and the Environment' (*The Ecologist*), 'It has been estimated that the operations of the armed forces may account for at least 6-10 per cent of global air pollution and that military-related activities may be responsible for 10-30 per cent of all global environmental degradation. Renner states that the total military-related carbon (dioxide) release in the US could be as high as 10 per cent. Furthermore, the armed forces of the world are the largest producers of hazardous chemicals and nuclear wastes'.

Bernard Lowen has written about the gulf war's oil fires, 'These fires spew noxious gases, soot and oil, mercilessly degrading the frail global eco-system. The burning wells and oil lakes released 50,000 tonnes of sulphur dioxide daily—a prime cause of acid rain—and 100,000 tonnes of sooty smoke. Meteorological satellites have detected smoke at a height of 6-7 kms and as far away as 2,000 kms from the source. Sunlight-absorbing plumes of smoke have been measured over an area of 42,000 square kms. The most dangerous aftermath relates to the release of greenhouse gases and sub-blocking soot production. The damage caused by the unhealthy smog and acid rain to trees, lakes, streams and rivers is incalculable and will continue long into the future'.

Hence, at a basic level, the concerns of these three leading movements are in conformity with each other. The conformity extends further to the movement for gender equality. The movement for equality cannot be complete without incorporating the equal socio-economic rights of women. As grassroots environmental movements have revealed time and again, rural women respond very well to the challenge of saving forests and water sources as their own lives are linked more closely with these and other natural resources. On the other hand, the protection of environment contributes greatly to the welfare of village women.

In modern war, the proportion of civilian victims is increasing steadily, and women increasingly bear the most burden. War refugees now have a very high percentage of women and children. In addition, in many wars, thousands of women become victims of rape and sexual violence. On the other hand, the compassion and emotive appeal which only women can bring can contribute greatly to the success of the peace movement.

The compassion that is inherent in all these movements will certainly also help the animal rights movement. The environment movement and peace movement will obviously make the habitats of innumerable species safer. On the other hand, protection of many animals will improve the productivity of farms and improve the livelihood of farmers and fisherfolk. Vegetarian diets will not only protect animals but in addition will help to reduce hunger by making more productive use of scarce food. Reducing the use of dangerous pesticides will improve the quality of food while also protecting several species which are not harmful.

While emphasising the harmony of these various movements and components of relevant social change, we are of course aware that temporary, short-term conflicts can also appear, but as long as the basic objective of genuine well-being of all is respected, these smaller conflicts can be resolved without disturbing the basic harmony.

As multiple crises of survival loom over us, it is truly a crucial time for all these movements to pool their energy and resources together to lay the foundations of a new, secure and happy world.

Peace Movement
Combine Global Issues with Local Concerns

AMIDST ALL the serious conflicts and many-sided violence of our world there is clearly need for well-organised, broad-based peace movements functioning with continuity. Instead, we've only smaller efforts dealing separately with various aspects of the agenda for peace. Bigger initiatives emerge only to meet specific challenges (such as the organisation of mass demonstrations against the invasion of Iraq in several western countries) for a short while. These efforts are good in themselves, but very inadequate.

There are many reasons why a strong worldwide peace movement (with close linkages of national chapters) is today more needed than ever before. First and foremost, of course, is the fact that the world has more destructive weapons in storage than ever before, capable of annihilating all human beings and most other forms of life in a matter of minutes. Possibilities of war between nuclear weapon states continue to exist. Increasingly, non-state actors (terrorists) are responsible for more and more violence, and their destructive capabilities are increasing. There is a real possibility of terrorists (who now control substantial territories) acquiring and using nuclear weapons, or other weapons of mass destruction.

The existing world leadership has only compounded problems in recent years by ill-advised actions, among other places, in Iraq, Syria, Libya, Sudan and Yemen. Entirely avoidable, highly dangerous situations have been allowed to develop such as the escalation of tensions between Russia and the USA. More and more countries are engulfed in bloody civil strife, often as a proxy or a sideshow of wider conflicts between bigger powers.

In this highly disturbing situation, an unbiased well-organised world peace movement can play a very important role in reducing distress and destruction but only if it is rooted in justice and truth while remaining free from all kinds of narrow interests. All kinds of powerful vested interests will almost certainly try to use the peace movement for their narrow ends, and so the movement should be strong enough with internal democracy and safeguards to keep away such pulls and pressures.

But apart from this, and this needs to be emphasised, the peace movement has a very important role in preventing the violence in our daily life. In fact, as long as WMDs are not used in wider conflicts, it is the violence in daily life which is much more destructive than the combined impact of all the external and internal wars in the world.

In an important report titled 'World Report on Violence and Health' (WRVH-prepared by the World Health Organisation) it is stated that in a single year 2000, 1.6 million persons died due to violence related causes. This included 310,000 deaths caused by war/civil strife, 520,000 homicides and 815,000 suicides. While the 310,000,10,000 deaths caused by war and civil strife are extremely sad, it is revealing to know that well over four times this number were caused by homicide and suicide.

The actual distress caused by violence in daily life is actually many times higher than what is indicated by the number of fatalities. According to WRVH, studies of non-fatal violence reveal that for every youth homicide there are 20 to 40 victims of non-fatal youth violence requiring hospital treatment. According to several experts, there are about 20 attempted suicides (in many cases leading to serious injuries) for every completed suicide.

In 48 population-based surveys from around the world, between 10% and 69% of women reported being physically assaulted by an intimate male partner at some point in their lives. For many of these women, physical assault was not an isolated event, but part of a continuing pattern of abusive behaviour. Studies from Australia, Canada, Israel, South Africa and the USA say that 40 to 70% of female murder victims were killed by their husbands or boyfriends, frequently in the context of an ongoing abusive relationship.

In a cross-sectional study of children in Egypt, 37% reported being beaten or tied up by their parents and 26% reported physical injuries such as fractures, loss of consciousness or permanent disability as a result of being beaten or tied up. In Ethiopia 21% of urban school children and 64% of rural school children reported bruises or swelling on their bodies resulting from parental punishment.

In the USA 44% of adolescent males in secondary schools reported involvement in physical fighting. In Canada a study revealed that 23% of girls had experienced sexual harassment while attending school. A study of health behaviour among school children in 27 countries revealed that the majority of 13-year olds in most countries had engaged in bullying. School teachers were found responsible for 32% of disclosed child rapes in South Africa.

In the European Union, an estimated 3 million workers have been subjected to physical violence at work. In the UK, 53% have suffered bullying at work. A lot of this violence leaves lasting impacts.

According to WRVH, several studies indicate a host of ailments (physical and mental) among women who have suffered physical or sexual abuse. Women abused by partners face an increased risk of suicide or suicide attempts. In Ethiopia several raped school girls are reported to have attempted suicide. In Sweden, workplace violence has been a factor in 10 to 15 per cent of suicides.

If the peace movement is closely involved in preventing and reducing all the forms of violence in our daily life, then clearly its potential to reduce human distress is greatly increased. This will bring the peace movement in close contact with a very large number of people on day-to-day basis. This mass contact can also be used to spread wider messages of world peace and prevention of war.

How war and violence in daily life can be related to each other is also brought out in some startling statistics presented in WRVH from Israel and Palestine. This report tells us

- 76 per cent of adolescent males in Jerusalem, Israel reported involvement in physical fighting in the past year.

- In Israel, in the under-18 age group, the annual incidence of violent injuries receiving emergency room treatment is 196 per 100,000 population.
- 52% of Palestinian women in the West Bank and Gaza Strip had been assaulted by their husbands in the previous 12 months.

In such conditions, there is enormous scope for peace movements to reduce violence and distress in everyday lives while also linking it with the reduction of the wider conflict.

By taking up issues relating to violence in daily life (in families, neighbourhoods, villages, schools, colleges and workplaces), the peace movement can reach out to all parts of the world and establish a close rapport with people. This enormous public support can then be mobilised for the wider causes of disarmament, including eliminating all WMDs, preventing war and setting up systems of peaceful resolution of all conflict situations.

Including Those Who Violate Peace

Peace is Important for All

EVERYONE REALISES that what our world needs above all is a message of peace. Yet violence exists on a very large scale and appears to be increasing at various levels—from violent tendencies in personal life to wars and conflicts at national and international levels. Violence as a means of removing the injustices caused by various kinds of violence still appears to appeal to many people.

In this difficult situation how to make the message of peace more effective is a challenging task.

- We can link up the distress caused by wars and the distress caused by violence in the personal lives in such ways that a very broad-based movement against violence or a movement for peace can emerge. Violence in personal life is generally rooted in the tendency to dominate others, to manipulate a relationship to one's advantage, to snatch the share of others or to use fair and unfair methods to get ahead of others. In a nutshell we may say that it is rooted in greed and dominance. But when greed and dominance become widespread in a society, then this society has a strong urge to grab the resources of other societies or to dominate other societies and for this may declare wars against them. So, there is a continuity between violence at various levels which causes enormous distress at various levels and it should be possible to build a broad-based movement which unites people against all this suffering as they understand that all these manifestations of violence are inter-linked.

- Secondly, it is important to realise that relationships of dominance harm people at both ends. On the one hand, the sufferings of those who are exploited are very clear. On the other hand it is also true that the group of people who exploit and plunder others for self-enrichment not only deprive others—in the process of plunder they also deprive themselves. The act of plunder and exploitation changes a person's soul in such a way that he becomes incapable of feeling and experiencing many kinds of happiness. Such a person tries to isolate himself as much as possible from the suffering caused by his actions, seeking also the help of high technology for this purpose, but such efforts are self-defeating.

  The biggest tragedy of this deprivation is that the loss is generally not realised by those who experience it, hence proper corrective action is seldom taken. In fact the persons (and societies made up of such persons) passing through this phase of deprivation are generally quite heady and arrogant due to a wrong sense of their power and dominance. Relationships of exploitation therefore destroy not only the exploited but also the exploiters. However, this tragic argument can also be seen at an optimistic level as efforts to end the exploitative relationships will liberate not just the exploited but also the (former) exploiters. A dominant characteristic of the exploiter is that he (it is generally he rather than she) doesn't want to be told about the self-defeating nature of such attitudes. An aggressive husband is just likely to dismiss the arguments given above. At the wider level of nations, the USA Government is prone to taking very strong action against other countries to counter the drugs problem but is unable to realise how the drug addiction problem in the U.S. society is rooted in its own spiritual deprivation. However, the efforts to bring such a realisation should continue as such a realisation on the part of exploiters is crucial to the creation of a better world.

Wars destroy not only the vanquished but also the victors, exploitation shatters not only the exploited but also the exploiters. The wealth obtained by a person or society cannot be separated from the methods used for obtaining this wealth. If the methods are unfair, if these involve the unfair deprivation of others (including future generations), then these methods also have in them the seeds of guilt that will prevent the person or the society from deriving real happiness on the basis of this wealth. To make this world safe and peaceful in the true sense of the word, the basic attitudes of greed, cruelty and exploitation have to be confronted and changed, instead of being hidden under hypocrisy, as these have the power to ultimately destroy not just the exploited, but also the exploiters.

Even though the economic, the political, the social, the ecological crises in the world are accentuating at a very rapid pace, the world's leaders, by and large, refuse to acknowledge that there is something fundamentally wrong with the world. If they talk of social justice at all, the suggestions they give are an insult to the basic changes that are needed. It is in this dismal situation that the above viewpoint—namely that social justice is essential to liberate not only the exploited but also the exploiters—should be seen. The increasing acceptance of this viewpoint, we believe, will help to create a public opinion that will not tolerate hypocrisy in the name of social justice, and thereby pave the way for real changes.

This relationship of 'burning at both ends' is also crucial in understanding high levels of depression and suicide, self-harm and self-abuse. In different ways, persons at both ends of a relationship of dominance and greed can be driven towards this.

- Lastly, to those who want to fight prevailing injustice by violence, we ask—is it not true that all this injustice has been caused in the first place by violence? If violence is the basic cause of most distress, can we really hope to reduce distress by ourselves embracing the basic cause of distress.

Beyond Material Goals

Genuine Spirituality Needed to Counter Hatred

IT WAS at one time hoped that spread of science and technology will lead to a reduction in religious fanaticism and the violence based on this. This has proved to be a false hope, as violence based on religious fundamentalism and fanaticism still remains a formidable and expanding force in the 21st century. This is often based on a firm belief in the superiority of one religion or one sect of a religion, and the willingness to use force to spread this religion (or sect of religion). Of course, hatred and contempt for other religions or sects is just the other side of this belief.

Among the various forces of this kind of narrow-minded and violent interpretation of religions, at present the Islamic State is almost certainly the most formidable and feared. However, such trends and forces exist in almost all the major religions to a lesser or greater extent. At various times these may be active or dormant, but these forces certainly exist in many countries.

To counter these forces of hatred, it is important to evolve a common spirituality relevant to our times which can be acceptable to people of all religions. In fact, such a spirituality linked in very rational ways to the real needs of our world can also be accepted by those persons who are not particularly keen about linking closely with any religion. Such spirituality can also be accepted by atheists and rationalists. As a first step, we need to have a common understanding of spirituality which is most relevant and useful for our society but is not linked in any narrow way to just one religion.

Spirituality is the sustained quest of human beings to be at peace with themselves by bringing their own welfare in harmony with the welfare of the entire world. This second aspect of linking one's own happiness to that of the entire world is of crucial importance and without this the concept of spiritual development is meaningless. A close look at the great religious and spiritual leaders of the world will reveal that this has been at the back of many of their teachings. Do not do to others anything that you do not want others to do to you, stay away from greed and violence, do not cheat others, extend all help to the poor and needy—these recurring themes (often hidden in fables and mythology) add up ultimately to the one broad precept of not causing any distress to others and reducing their distress whenever possible. In other words, bringing our welfare in conformity with the welfare of the world.

Spiritual development is our sustained quest for getting rid of value systems which create unnecessary, entirely avoidable stresses and replacing them with values which expand our horizons of happiness and take us to a higher level of happiness that cannot be achieved from materialistic accumulation of riches or relationships of dominance. However, as emphasised earlier, this spiritual development of the self is complete and meaningful only when it is related to the welfare of other people and other forms of life. Achievement of inner peace or deeper knowledge is not an end in itself—it becomes invaluable only when it is related to the efforts for creating a better, happier world. Therefore, spiritual development also demands that as one gains a fuller understanding of what a better society is, so efforts are also made to tie up with the larger groups and forces that are trying to create a better society.

In this definition, it is important that this quest should be sustained. The most essential feature of this sustained quest is that we bring our own welfare in harmony with the welfare of the entire world. In practical life, what this means is that in all our actions and thinking, we keep the welfare of other people and other forms of life in mind. We'll never like to promote our interests in ways that harm other people or other forms of life. We'll like to find such a path of life where we can integrate our own welfare with the welfare of others.

This path of spirituality can be followed by persons belonging to any religion, by secular persons or even atheists. People having firm faith in their religion can gather strength from their own religion to remain firm on the path of spirituality.

At a time of growing sectarianism and clashes between various religions, a path of spirituality can be evolved as a common heritage of all religions. People having firm faith in their own religion can use this common path of spirituality as a bond of unity with other religions, as a meeting place where all religions converge on some basic principles of how we can create a world of justice, peace and harmony. This is as close as one can hope to come to the concept of a universal religion.

Apart from creating spaces for countering the culture of hatred, the spread of such genuine and rational spirituality will also help to create favourable value systems for justice and equality, environment protection and peace.

When messages of environment protection, peace and poverty eradication are preached in the present-day world, these quite often fall on deaf ears. But it is different with those people who have accepted the path of genuine spiritual development. Their life is already in conformity with these objectives, and they are prepared to make any further changes and take any further initiatives that are needed for the welfare of all and for the protection of environment. Actually spiritual development is nothing but the voluntary, happy and enlightened acceptance of a personal lifestyle which is fully in conformity with the welfare of the entire world, all its life forms and ecosystems.

For a Better World

Convergence of Social Justice and Spirituality Movement

IT IS very widely agreed that intolerably high levels of economic inequalities and social injustices exist in India and these are a very important cause of distress and deprivation in our country. There are several sincere social movements which have been struggling against these inequalities at various levels. The highly relevant role of those which are sincere is widely acknowledged. These movements on the one hand seek some relief for deprived people and victims of injustice and on the other hand seek to create a much more equal and just world where the possibilities of deprivation and the related distress will be greatly reduced due to equality on a longer-term basis.

India is a land which has been known to be rich in several sincere spiritual movements. These movements generally have been concerned with the development of high character of individuals (or groups of individuals) and providing them a firm base of value systems so that they can remain free from greed, dishonesty, arrogance , dominance, excess of sensual pleasures, various addictions, overwhelming personal ambition, violence and falsehoods while remaining close to truth, honesty, love, peace, cooperation and a life of simplicity based on limited needs. Any movement which helps to create more and more such individuals helps to create a better world in which distress will be reduced at various levels in various ways.

Thus at a most basic level those movements of social justice which are sincere and those movements of spirituality which are sincere have one basic similarity—they aim to create a better society and reduce distress . It is also most likely that as long as they are sincere, both will succeed in achieving this aim to a lesser or greater extent.

However, beyond this very basic commonality there are many other possibilities of convergence and mutual support which can increase the impact of both movements for creating a better world.

A movement for social and economic justice can succeed in creating a more equal society by making structural changes in economy but in order to sustain these gains it will need spiritual development. If by changing ownership of means of production you create a more equal society for the time being but people and leaders have the same weaknesses of greed, lust, arrogance, addictions, dishonesty and dominance etc. and no efforts are made to check these then after some time new problems and tensions including trends towards new forms of inequalities, are bound to appear again. Ecological ruin will also increase which in the present-day world is likely to be most harmful.

On the other hand, if a movement for social justice after

its initial success in creating a more successful society also emphasises spiritual development adequately to improve the most basic virtues of truth, honesty, simplicity and limited needs, cooperation and non-violence, then the possibilities of maintaining equality, justice and peace as well as ensuring broad-based and sustainable progress based on cooperation of all will be significantly improved.

Also, it is very important for the movement of social justice to maintain the high commitment and dedication of its members over a long period of time and, for this, effort of spiritual development and an overall environment of spiritual progress influenced by sincere spiritual movements is very useful.

On the other hand, for the movement of spirituality also, the movement for social justice plays a very helpful role as the development of spirituality is much more possible in overall conditions of justice and equality than in conditions of inequality and injustice.

Imagine a village in which landlords exploit others almost as bonded workers. Then if someone who wants to create a better world goes to bonded workers and without talking about ending their exploitation talks only about honesty and truth to them then this will appear ridiculous. On the other hand, if exploitation is being challenged at the same time then it becomes much more sensible to talk about spiritual development as important in itself and as being very helpful for the struggle of workers. Or imagine a situation in which exploitation has already been checked to a considerable extent but the gains of betterment have to be stabilized and here again the role of spiritual development becomes very important to complement and complete the earlier achievement of the movement for social justice.

Thus, while the movement for social justice added strength

alongside and in cooperation with the movement for spiritual development, in isolation the movement loses some of its strength and potential for wider benefits to society. Imagine a big company involved in exploiting farmers and workers which holds meditation classes for its officials to improve their mental health. A question that arises is—is this to improve their health for increasing their competence to exploit others?

Such contradictions do not arise when social justice and spirituality work to complement each other. These are essentially two very beautiful branches of a tree which when they converge and complement, the beauty and fragrance of both branches is further enhanced. Sincere people belonging to both these movements need to come closer to each other and explore possibilities of closer cooperation.

The Essence of History

It is all about Four Crucial Relationships

IT IS increasingly realised that instead of concentrating on kings and queens, conquests and invasions, we should concentrate on how the lives of ordinary people—farmers, workers, artisans and others—have changed over the years. This is certainly needed, but beyond this there is a real need to further widen the scope of our concerns. We need to look at the welfare of not just all ordinary human beings but all life-forms—whether birds or mammals, reptiles or fish, whether farm animals or wildlife or pets. This makes our task bigger, more ambitious and more difficult, but the only way to get a real idea of how life on earth has changed is to look at how all life-forms have changed.

But there is no doubt that it's the human species which has a unique capacity (used for good or bad is another matter) to influence life on earth and this is why the relationship of human beings with other life forms needs to be studied closely.

What we should try to learn from history above all is what were the most important periods of closer and better cooperation among human beings for the welfare of all? What were the times when people were most compassionate towards each other and towards other forms of life? What were the factors which made such achievements possible and what were the factors that contributed the most to it?

The history of earth since the advent of humankind on it can be largely understood as the history of four crucial relationships—the relationships among various human beings at various levels, the relationship of human beings with nature, the relationship of human beings with other forms of life, and the relationship of the present generation to the generations to come.

Generally, the welfare and well-being of our world have been advanced and distress has been reduced when these relationships were formed on the basis of cooperation, care and harmony. On the other hand, distress and destruction have increased when these relationships have been based on dominance.

Relationships among Human Beings

If only one cause has to be singled out for the prevalence of large scale distress in the world, then this is likely to be the persistence of relations of dominance. It has been a very widespread and enduring practice of human beings to try to dominate others, to get personal benefit at the cost of others, to try to get ahead of others and to impose their own viewpoint. This tendency may be rooted in greed (the urge to get more than others, regardless of one's needs) or in ego (the urge to impose one's own thinking and viewpoints, without giving a patient and fair hearing to the other viewpoint), but its impact is always harmful. This tendency can be seen in the relationships of individuals, groups, classes and entire nations, as also in gender-relationships.

This relationship of dominance is also responsible for the most exploitative economic systems and the most cruel wars which at their worst have claimed millions of lives.

The political history of humankind since ancient times is largely dominated by the strong urge to conquer other nations and people. Genghis Khan, the thirteenth century Mongol invader is believed to have said, 'A man's highest job in life is to break his enemies, to drive them before him, to take from them all the things that have been theirs, to hear the weeping of those who cherished them, to take their horses between his knees, and to press in his arms the most desirable of their women'.

While earlier wars witnessed much cruelty and distress, the scale of this was limited by the technology available at that time. From 16th century onwards new discoveries and inventions opened up the entire world to invaders, and in addition armed them increasingly with weapons of mass destruction.

The French philosopher Montaigne wrote at the end of the sixteenth century, 'So many goodly cities ransacked and razed; so many nations destroyed and made desolate; infinite millions of harmless people of all sexes, states and ages, massacred, ravaged and put to the sword; and the richest, the fairest and the best part of the world topsy-turvied, ruined and defaced for the traffic of pearls and pepper'.

Las Casas who was an eyewitness to the Spanish conquest of South America described the genocide of the native American population in Espanola (present-day Haiti and the Dominican Republic) as follows:

> As they rode along, their lances were pierced into women and children, and no greater pastime was practised by them than wagering as to a cavalier's ability to completely cleave a man with one dexterous blow of his sword. A score would fall before one would drop in the divided parts essential to winning the wager. No card or dice afforded equal sport. Another knight from Spain must severe his victim's head from the shoulder at the first sweep of his sword. Fortunes were lost on the ability of a swordsman to run an Indian through the body at a designated spot. Children were snatched from their mother's arms and dashed against the rocks as they passed. Other children they threw into the water that the mothers might witness their drowning struggles. Babes were snatched from their mother's breasts, and a brave Spaniard's strength was tested by his ability to tear an infant into two pieces by pulling apart its tiny legs. And the pieces of the babe were then given to the hounds that in their hunting they might be the more eager to catch their prey. The pedigree of a Spanish blood-hound had nothing prouder in its record than the credit of half a thousand dead or mangled Indians. Some natives they hung on gibbets, and it was their reverential custom to gather at a time sufficient victims to hang thirteen in a row. I have been an eye-witness of all these cruelties, and an infinite number of others which I pass over in silence.

This cruel behavior, at least for some time, had the support in the home countries at the highest levels:

> In 1775 King George III of Britain in a Royal Proclamation decreed that 'For every scalp of a male Indian brought in as evidence of their being killed... forty pounds... for every scalp of such female Indian or male Indian under the age of 12 years... twenty pounds'.

Third World Resurgence magazine has written:

> Indian generosity and honour were repaid with European treachery and cruelty. The whites broke every treaty they made with the Indians to deprive them of their lands. Between 1887 to 1934 the Indians lost some 100 million acres of their land to the white man. Nations of Indians were destroyed and European diseases were used to exterminate them.

In the Second World War, according to some estimates nearly 75% of the Jews living in Europe were killed. Nearly one-fifth of the entire population of Poland was killed while nearly one—tenth of the entire population of Soviet Union was killed.

In the post—war years, several countries/regions have lost more than one tenth of their populations in war and civil strife and/or liberation struggles of Vietnam, Korea, Iraq, Rwanda, Burundi, Algeria, Angola, Sudan, Afghanistan, Cambodia, Timor, Tibet and perhaps a few others. In some countries like Combodia and Rwanda it will be nearer to the truth to say that one of five have been killed.

In the 20th century, the two world wars and other major conflicts have claimed nearly 100 million human lives, at the average of nearly 1 million lives every year. In the last decade of this century the world has accumulated nuclear stockpiles which add up to a destructive potential which is nearly 700 times that of all the explosive power used in the three major wars of this century, enough to kill all human beings (as well as most other forms of life) many time over.

The relationships of dominance and exploitation of course destroy the dominated and exploited people. But in addition, and this should be emphasised, they also slowly but surely destroy the perpetrators of exploitation and domination.

As I've written elsewhere, 'The persons who fill their coffers by inflicting injustice and cruelty on others have to live with a guilt complex that can destroy their peace. To get rid of this guilt they have to lower themselves to such a level of insensitivity that deprives them of simple yet precious joys of life. The suppressing of spiritual values (that is required to be able to commit injustice) creates distortions in the persons (or the communities) due to which they are denied even that happiness which many poor people can experience freely and in plenty'.

In other words, a person (or a group or a nation) who inflicts injustice and injury on others, will either live with a guilt complex (if he wants to retain some sensitivity) or else he'll have to reduce himself to a level of insensitivity that will prevent him from feeling small but precious joys of everyday life; this in turn is bound to adversely affect his closest relationships including those with his family members. Thus, relationships of dominance are not only destructive, these are also self-destructive.

It is true that efforts have been made to use high technology in such ways as to create a distance between the perpetrator of injury and the effects of his actions in such a way as to remove the feeling of guilt. Ravi Sundaram writes,

> The emergence of a complex division of labour under modern capitalism has meant that functional specialisation generates a necessary remoteness of human agents from the end-product of their social action. In this context, the bureaucrat's own action becomes an end in itself. Once so isolated from the consequences of action, the bureaucrat, untroubled by moral dilemmas, can pursue his allocated tasks. The architects of the holocaust, the bombers of Hiroshima and Nagasaki, and the technocrats who designed the Vietnam war, could continue doing

> so without moral qualms precisely because of the social production of distance in modernity. This aspect is crucial in understanding the technological evolution of the holocaust. In the early history of the holocaust, the victims were rounded up and machine-gunned at point-blank range. The administrator soon found this both primitive and inefficient - and damaging to soldiers' morale. Other techniques were sought which would preserve the optical distance between murderers and murdered. The result—the gas chamber, the perfect murdering machine. This reduced the role of the killer to that of the 'sanitation officer' who simply pressed the button which released the gas into the chamber filled with the hapless victims.

However, technological solutions to remove guilt and related afflictions have not always worked. *Reader's Digest* has described the suffering of a child, Kim Phuc at the time of bombing of Trang Bang (Vietnam) by US planes (RD November 1997):

> 'The bombs, canisters filled with napalm, had smashed into the ground behind Kim and instantly ignited. The jellified gasoline, designed to stick to and incinerate anything it touches, splashed onto Kim's back. Her flowered cotton shirt and pants–even her sandal–combusted. She was engulfed in a cloud of smoke and fire as napalm peeled away the skin from her back and left arm. Terrified, Kim kept running. At first she could feel nothing. Then she felt as if she had been thrown onto an open fire. In horror she saw the skin drop off her arm like clothes off a doll. As she ran naked down the road that led out of the village, she began screaming, 'Too hot! Too hot! Please help!'

This magazine also noted the impact of this suffering on the sensitive mind of the pilot who caused this suffering:

> Now he stared at the picture of Kim Phuc, her agony caught for eternity. His own son Louis was about the same age. He could almost smell the child's burning flesh.
>
> ...Later he kept his role in the bombing of Trang Bang secret, locked deep within his soul. It surfaced in the form of a nightmare. First Plummer would see a picture of Kim, with arms outstretched and mouth frozen in a silent scream. Then the image would widen to include Kim's brother and cousins running alongside her. Finally, he would hear their screams, louder and louder until he felt surrounded by the accusing children. To drown his guilt, Plummer began drinking heavily. In July 1973 he married for the second time, but he still kept his secret. No one can understand, he thought. John Plummer's drinking cost him his marriage in 1979. It was a vicious circle; he drank to put the bombing out of his mind, but the drinking made him more obsessed.

Time magazine reported recently that spousal abuse is occurring in 1 out of every 3 US Army families each year—double the civilian rate. Some cases of violence at the hands of husbands (or other men in family) are so gory that not many people will be able to stand any detailed descriptions. As reported in *Time*, a marine in California, who was a 'hero' of the Persian Gulf war, not only killed his newly-divorced wife but in addition also killed their five-year-old daughter. In North Carolina, an airman hacked his wife to pieces, wrapped her remains in plastic garbage bags and stored them in the refrigerator.

If A exploits B, then B suffers but directly or indirectly, A also suffers. Another frequently-seen aspect of relationships of dominance is that instead of confronting A, B in turn tends to exploit C (being someone over whom he can easily exercise his control or vent his frustration). For example, workers faced by retrenchment or other injustice at worksite sometimes resort to domestic violence. In joint families in India and elsewhere women facing domination take it out on other women members of the family.

Extremely tragic as all this is, an understanding of these relationships of dominance also opens up possibilities of convincing more and more people about this 'Burning at both ends' (this was also the title of my book on this subject)—as the process of exploitation and dominance has within it seeds of destruction even for the exploiter, therefore, it'll eventually help everyone to try to re-build the world on the foundations of relationships of cooperation and harmony.

Secondly, it is important to identify those periods of history when the relationships of equality, harmony and cooperation were more in evidence, and to try to build further on these sources of inspiration. We need to know more about the forces, movements and personalities who helped to bring such periods based on harmony, peace and cooperation, so that we can learn from such efforts.

As the study of such times will tell us, relationships of dominance may have been more frequent in our history, but these are by no means inherent in human nature. There is a lot of evidence today of gender-relationships being affected very adversely by dominance and the injustice arising from it.

The Human Development Report (HDR) 1995 summarised the existing studies on the continuing discrimination against women. Some important information that emerges from this report is given below:

The number of illiterate women is double that of illiterate men. Women suffer more than men from malnutrition. Near

500,000 maternal deaths take place every year in developing countries, including nearly 330,000 in Asia and 150,000 in Africa. Wage-rates are generally lower for women. Although women constitute half the electorate, they hold only 10% of the seats in the world's parliaments and 6% in national cabinets. There is widespread discrimination against women in health, nutrition and at other levels, to the extent more than 100 million women are 'missing'.

According to HDR (1995), studies in four developing countries suggest/indicate that two-thirds or more of married women have suffered domestic violence. In the USA domestic violence is the biggest single cause of injury to women. Further studies show that a third of women in Barbados, Canada, the Netherlands, New Zealand, Norway and the USA report sexual abuse during childhood or adolescence. Each year an estimated one million children, mostly girls in Asia, are forced into prostitution. Studies in four developed countries suggest one woman in six is raped in her lifetime.

But inequality, domination and discrimination was not always the basis of gender relationships. In fact if we consider the roughly 40,000 years of the history of human beings on earth, then probably for the first 30,000 years or so the gender relationships were more equal and just in most parts of the world. As Chris Brazier writes:

> In fact the early people would have eaten much more vegetable than animal matter—it was women who gathered the fruits, nuts and berries that sustained people most of the time. And if hunting and gathering people that still exist are anything to go by, these early humans probably had a deep respect for women and their contribution. ...Women could be counselors, leaders, doctors and law

> givers—and had their own special status as holders of the most sacred mystery of all—that of birth… From 25,000 BC the worship of the Great Goddess Mother was practised all over the world, from the steppes of southern Russia to Australia. Women's status was much more equal then it is today.

As an example Brazier mentions that in the marriage ceremony of ancient Egypt, the husband pledged never to oppose his wife's wishes.

This shows that human relationships of dominance, inequality and injustice should not be taken for granted. Few people know that for the greater part of human history, gender relationships were more equal and harmonious than what these have been in recent decades and centuries. Such aspects of history need to be better known or else we can easily become too pessimistic.

If we take examples from India's history, then despite all the conquests and invasions with which we are all too familiar, there were several periods when those who preached the message of harmony, cooperation, equality and simple, non-violent, truthful life attracted very large numbers of people.

Hence for a very large number of people for a considerable time, life based on these precepts was accepted as the true life. India and its neighbourhood (including present-day Nepal) around 6th century BC was one such place where the message of Gautam Buddha, Mahavira and other sages spread far and wide. Such views also flowered during the reign of King Ashoka around the 2nd century BC. During medieval times, again we see a very inspiring revival of such ideals in the form of the teachings of Guru Nanak, Sant Kabir, Garibnawaz and other saints of Bhakti and Sufi movements. The freedom movement is another inspiring time when a very large number of people made tremendous sacrifices to fight injustice and dominance.

Relationship of Human Beings and Nature

Relations of dominance are seen not only among human beings but also between man and nature and between man and other forms of life. The tendency of looking upon nature as something to be dominated and conquered has been responsible to a significant extent for ecological ruin. Of course, human beings have to meet their various needs, obtain food, clothing and shelter, and in the process they have to make certain demands on nature—they have to obtain water from rivers and food from land. It is certainly possible to do so while maintaining an attitude of respect and coexistence towards nature, preserving the clean and beautiful flow of rivers and protecting the fertility of land. There is evidence that in some ancient cultures there was an attitude of reverence towards nature, an attitude which survived till much later times among many indigenous groups.

Chris Brazier has summarised the teaching of Lao-tzu (Southern China around 600 BC), 'The gist of his teaching was that life and nature form an indivisible whole, with every positive element having its equivalent negative aspect, every *yang* its *yin*. According to his belief, the earth and the universe regulate themselves according to certain eternal principles; this was the Tao or the way, and the religion based on his work thus became known as Taoism or Daoism. Campaigners for ecological concerns in the West in the last few years have noticed the similarity between their notion of the planet as a self-regulating ecosystem which humans have abused and the Taoist conception of life on earth.

> Providing a summary of Mayan ethics the *New Internationalist* writes, 'According to the Guatemalan Mayan vision of the cosmos, every form of life emerges from the same origin or seed. Some seeds become trees, others flowers, others water, others human beings. Thus, each creature is inextricably linked to all others and what one does to a tree affects not only the tree but oneself and other creatures. This inter-relatedness calls for profound respect between people and their Creator, between people and nature, and among people themselves. The aim of the Maya is to keep their relationships with the world around them, and also the inner life of each person, in perfect balance according to the rhythms of the cosmos'.

This journal goes on to say that Mayan ideas have much in common with those of other indigenous cultures of the Americas, especially in their holism and respect for the environment.

In 1855, the Indian Chief of Seattle responding to pressures from the United States President to sell the land of what is now Washington State, had this to say,

> How can we buy or sell the sky or the warmth of the land? Such thoughts to us are inconceivable. We are not in possession of the freshness of the air, or the water-bubbles. Every corner of this land is holy to my people. They remain holy in the memory of my people—from the sparkling pine leaves, the sandy beaches and the mist of dark brooding forests, to the songs of insects... We know that white men do not understand our way of life. Land to him is not a brother but an enemy. After conquering a piece he proceeds to the next... Our God is the same God that you worship. His compassion extends equally to white men and Indians. This land is precious to Him and harming it, therefore, would be an insult to our Creator.

However, these views of nature increasingly came in conflict with the tendency which existed even in ancient times, of making excessive demands on nature, inflicting grave damage on land and water sources, and thereby sooner or later also bringing disaster on human beings. John Bellany Foster writes, 'The history of precapitalist and preindustrial societies is full of examples of social collapse brought on by environmental depredations'. Thus, while attitude of reverence towards nature certainly existed in ancient and indigenous traditions, they should not be romanticised too much as the conflict with other viewpoints based on excessive expansion and exploitation appeared sooner or later in most places.

In the conflict of these views—one stressing conquest of nature and the other stressing coexistence with nature—the former attitude started asserting itself more and more with the passage of time. The progress of science should have opened our eyes to the dangers of making excessive demands on nature but in reality something entirely different happened. The unravelling of the mysteries of nature appears to have decreased the awe of it and encouraged the view that as we know its secrets we can conquer and dominate it. Philosopher of science and one-time Lord-Chancellor of England Sir Francis Bacon observed that the conquest of nature constitutes, 'the real business and fortune of the human race'. He said nature must be 'bound into service' and made a 'slave'.

Such a viewpoint cleared the way for and provided the justification for very large-scale disruption of environment in the last few centuries. However, as rivers were turned into sewer-lines, the rain became acidic, and even the life-giving sunshine was made hazardous by the depletion of the ozone layer, during the last few decades there has been growing realisation of the need for harmonious coexistence with nature instead of striving to dominate it. Domination brings destruction while a protective attitude towards nature also protects the life and livelihood of people.

Engels wrote,

> The people who, in Mesopotamia, Greece, Asia Minor, and elsewhere destroyed the forests to obtain cultivable land, never dreamed that they were laying the basis for the present devastated condition of these countries, by removing along with the forest the collecting centres and reservoirs of moisture. When on the southern slopes of the mountains, the Italians of the Alps used up the pine forests so carefully cherished on the northern slopes, they had no inkling that by doing so they were cutting at the roots of the dairy industry in their region; they had still less inkling that they were thereby depriving their mountain springs of water for the greater part of the year, with the effect that these would be able to pour still more furious flood torrents on the plains during the rainy season... Thus at every step we are reminded that we by no means rule over nature like a conqueror over a foreign people, like someone standing outside nature—but that we, with flesh, blood, and brain, belong to nature, and exist in its midst, and that all our mastery of it consists in the fact that we have the advantage over all other beings of being able to know and correctly apply its laws.

Relationship of Humankind with Other Forms of Life

As the most capable and powerful form of life on earth, human beings have a special protective role towards other forms of life, but instead they've adopted a highly destructive role.

We'll also do well to remember that all other forms of life, whether in the oceans or on land, whether wild or domesticated, have to bear the adverse impacts of the environmental ruin unleashed by us, while we human beings alone are responsible for unleashing this ruin. When our activities bring death and disease to them, we do not even hear their silent protests.

Harvard Professor Edward O Wilson, one of the world's leading experts on biodiversity recently summarised the current state of other forms of life in an article in *Time* magazine. Biologists generally agree, he says, that, 'on the land at least and on a worldwide basis, species are vanishing 100 times faster than before the arrival of Homo Sapiens'.

> The ongoing loss in biodiversity is the greatest since the end of the Mesozoic era 65 million years age. At that time, by current scientific consensus, the impact of one or more giant meteorites darkened the atmosphere, altered much of earth's climate and extinguished the dinosaurs. Thus, began the next stage of evolution, the Cenozoic era or age of Mammals. The extinction spasm we are now inflicting can be moderated if we choose. If not, the next century will see the closing of the Cenozoic era and the start of a new one characterized by biological impoverishment. It might appropriately be called the Eremozoic era, the age of loneliness.

It is significant that half of the world's forest loss over the course of human history occurred between 1950 and 1990. In the fragile ecosystems of the world's tropical forests—where 50 per cent of the world's species reside—half of the forest cover is now gone and half of what remains is fragmented and degraded.

Afforestation drives have led to new tree cover but this is mainly in the form of tree plantation. As JB Foster writes in *The Vulnerable Planet*:

> Industrial tree plantations are biological and genetic deserts when compared to the rich complexity of natural forests. The streams that flow through these plantations contain few fish. The variety of plants, animals, insects and fungi is minimized . The floor of an old-growth forest is a lush carpet of vegetation; the floor of a tree plantation is almost barren by comparison. The trees themselves, which are viewed as mere commodities (i.e. so many board feet of standing timber), are 'genetically improved' to allow for a lower rotation time and hence for profit maximization. Natural diversity is destroyed in the same proportion as profits are promoted.

Accordingly to FAO's recent detailed studies (as reported in WRR 1994-95) tropical forest areas have been shrinking on average 15.4 million hectares per year. A recent analysis (quoted in HDR 1994) of tropical forest habitats which contain 50 to 90 per cent of the world's species, concluded that, at current rates of loss, up to 15% of the earth species could disappear over the next 25 years.

Some of the most 'developed' countries have also been the most ruthless towards other forms of life. As World Resources Report (WRR 1990) says, 'Nowhere has the conversion of natural habitats been more ubiquitous than in the temperate portions of Europe, where essentially no natural habitat remains. Wildlife still exists on the continent, of course, but the number of extinctions is significant'.

According to WRR 1992-93, 'Recorded extinctions over the past 400 years indicate that 56 mammal species and 115 bird species have disappeared, most of them on islands. About 12% of mammal species and 11% of bird species were classified as threatened in 1990'.

Not just tiny, vulnerable creatures but even giant, mighty

species are falling prey to the arrogant exercise of their unlimited power by human beings. The number of blue whales has come down to 10,000 from an estimated pre-whaling population of 300,000. The black rhinoceros population was reduced from 65,000 in 1970 to under 3,000 in 1990. In several countries more than one-third of the total known species (of mammals and birds in some countries, reptiles and amphibians in others) are threatened. Even several species of insects like bees and butterflies face extinction in several places.

New International magazine has summarized some important data on depletion of wildlife—100 million animals are killed every year for their fur. There were 100,000 tigers at the turn of the century of which less than 8,000 now remain in the wild. The world's elephant population has declined by 60 per cent in a single decade. Over 1,000 animal species are under threat of extinction, thanks largely to human action.

According to J B Foster:

> The rise of capitalism led to a loss of wildlife on a scale never before seen in human history. The spread of commerce resulted in the death of hundreds of millions of large animals at the hands of traders. In the fifteenth century, sables were common as far west as Finland, but by the late seventeenth century they could only be found in Siberia. By the end of the eighteenth century, nearly every species of fur-bearing animal in Siberia had been decimated and Russian fur traders had to move to the northern Pacific islands, where they killed 250,000 sea otters between 1750 and 1790. With the extermination of fur-bearing animals in western Russia, the fur trade became one of the driving forces behind European expansion into the North America. In 1743 the French port of La Rochelle, a centre of trade with Canada, imported the skins of 127,000

> beavers, 30,000 martens, 1,200 wolves, 12,000 otters and fishers, 110,000 raccoons and 16,000 bears.

Western Ecuador is reputed to have once contained between 8,000 and 10,000 plant species, some 40 and 60 per cent of them endemic. Given that there are between 10 and 30 animals species for every plant species in similar areas, western Ecuador must have contained about 200,000 species. Since 1960, almost all the forests of western Ecuador have been destroyed to make way for banana plantations, oil wells and human settlements. The number of species thus eliminated is difficult to judge, but the total could well number 50,000 or more, all in just 25 years.

According to UNEP, 'Around one-third of the world's coastal regions are at high risk of degradation, particularly from land-based sources of pollution and infrastructure development. European coasts are most affected, with some 80% at risk, followed by Asia and the Pacific, with 70% of the coast at risk'.

Although serious local threats to the biodiversity of oceans and seas have been reported for a long time, recent updated estimates of the extent of the loss to world fisheries have set alarm bells ringing. In a recent paper published in *Nature* journal, two Canadian biologists Ransom Myers and Boris Worm, have presented estimates which show that in the last 50 years, overfishing has removed nine out of ten large predators—big fish like tuna and cod. These latest and most comprehensive estimates by biologists from Dalhousie University have imparted greater relevance and significance to several scattered reports which had warned earlier about the fast depleting fisheries and other forms of life in several ocean areas without attracting much attention.

The WRR(1991) says:

> Many regional fisheries in fishing areas such as the NorthWest Pacific and the Southern Ocean show signs of drastic over fishing... The Antarctic fisheries turned from one species to another during the past 18 years as each harvested species has declined because of overfishing.

According to another estimate published in September 1997, populations of such species as cod, hake, haddock and flounder, 'the seemingly inexhaustible mainstay of North Atlantic fisheries for four centuries', fell by as much as 95% between 1989 and 1994.

How did such enormous losses occur? Largely because legally permissible fishing technology had become 'as devastating as dynamite' and in addition, several illegal practices were also adopted. The permitted technology included computerised ships as large as football fields and nets wide enough to swallow a dozen Boeing 747s which could gather up 120 metric tons of fish in a single sweep.

Describing the use of this technology, the acclaimed report 'Imperilled Planet' published by MIT wrote: Every evening as the Sun goes down over the deep blue of the Pacific Ocean, thousands upon thousands of km of almost invisible nylon netting are unfurled into the sea. Besides their intended catches of squid and tuna, these almost indestructible walls of death, each up to 65 km long and 12m deep, trap hundreds of thousands of dolphins, seals, turtles, sharks, salmon and seabirds in an indiscriminate marine holocaust.

According to another estimate, in 1993, shrimp trawlers in the Gulf of Mexico caught and threw away an estimated 34 million red snappers, including many juveniles. High technology made it possible to sweep up in a few days catches that earlier took a season, even though much of the catch was dumped overboard or used later as animal feed or even fertiliser.

Apart from overfishing, pollution too has contributed much to the alarming loss of biodiversity in oceans and seas. According to estimates for a single oil spill disaster, the Exxon Valdoz in 1989, 350,000 sea birds and millions of marine creatures perished in this disaster. Damage was so serious as to afflict generations of some species from the toxic and carcinogenic components of pollutants. In view of the great distress that a single oil spill can cause, their frequency is really disturbing.

Zhores A Medvedev writes about the adverse effects of dams on Volga river:

> The ecological problems caused by the Volga 'cascade' (as the series of dams has come to be known) have developed gradually. Many of them could have been predicted if the planners had taken environmental considerations into account. Eighty five per cent of the world production of sturgeons is found in the Caspian-Volga system, and it should have been obvious that the construction of the Volgograde dam in the lower part of the Volga would damage fishing by interrupting the annual migration of Caspian fish to their upstream spawning grounds. Artificial fish nurseries were installed, but they interrupted the natural migration patterns of the fish, with disastrous results.
>
> Pollution and the loss of oxygen in stagnant reservoirs added to the problem. Massive fish kills have occurred every year from industrial pollution and toxic waste. Pollution is also thought to be responsible for damage to fish. In 1984, 20 per cent of the sturgeon stock was found to be suffering from muscular dystrophy, causing a

> deficiency of eggs (caviar) in female sturgeon. By 1987, 90 per cent of all sturgeon species had developed the disease and in 1989 the whole population was affected. The same illness has also affected beluga, carp and herring.

One would have hoped that at least their nearness and daily usefulness would have led to a somewhat better attitude towards farm animals; in the village communities of many countries like India there are many examples of farm animals, particularly cows and bullocks being tended affectionately almost like family members. However, the uprooting of traditional farming practices and the drift towards factory farming has been accompanied by growing insensitivity towards the feelings of animals. In fact, from some points of view the more modernised, mechanised and productive the factory farms become, the plight of the animals gets steadily worse.

In factory farming, to an amazing extent there is a single minded pursuit of increasing production while completely ignoring the distress of animals. Chickens are placed so close to each other that they cannot even move, and their beaks are cut off to prevent them from pecking each other in frustration. By nature scrupulously clean animals are forced to remain for long hours in their own excreta. The reckless pursuit of quick and fleshy growth of animals to get more profits leads to equally reckless search for cheap sources of high protein feed and animals are sometimes given feed based on their own waste and the meat of their own brothers and sisters (animals and birds of the same species). Cows are routinely given oxytocin injections to pull out even that milk which they are not yet ready to yield, causing them a lot of pain and harming their health. Indiscriminate exposure to chemicals and antibiotics has become widespread.

The extent to which farming practices have been dehumanised can be seen from the following advice given by a leading magazine of 'pig production engineers' called *Hog Farm Management*: Forget the pig is an animal. Treat him like a machine in a factory. Schedule treatment like you would lubrication.

Some time back when the price of wool fell in the international market millions of sheep were killed in Australia and mass graves were dug to bury them. *Time* magazine reported on January 14, 1991:

> Across the rolling countryside, the normal peace of rural life is being shattered these days by volley of gunfire. Under the hot summer sun in the Southern Hemisphere, rifle shots echo over the rangelands as sheep farmers execute one of the largest animal holocausts in history. The scenes can be pitiful. Some farming families turn away in tears and drive off quickly after delivering their gentle charges to the slaughtering pens. There, next to mass-burial pits dug in grazing tracts or on the outskirts of rural towns, firing squads take over in a project designed to kill 20 million sheep during the next year.
>
> The death sentence was decreed as an emergency measure to curtail wool production and rescue a vital export industry. Over the past year-and-a-half, Australia's prime overseas customers have cut back on purchases, leaving a glut of fleeces on the market. Beyond the wool crisis, wheat farmers expect to see their incomes halved this year, and homegrown-citrus sales have also soured. At a time when much of Australia is taking to beaches and playgrounds, high summer in the Lucky Country's interior has become a nightmare.

If anything worse is possible then this is inflicted on animals used for various laboratory experiments and testing for products like cosmetics and drugs. *New Internationalist* has summarised this pain and suffering. Nearly 200 million animals are subjected to painful experiments in laboratories around the world, even though medical research on animals has been judged to be 71 per cent unreliable. In the course of these tests, animals may be given electric shocks, injected with diseases or experimental drugs subjected to toxic chemicals or radiation.

Thus, we've been very selfish and increasingly very cruel in our attitudes towards other forms of life—in farms and forests, in rivers and oceans. It is one of the most compelling requirements of ethical life that this attitude should change in a basic way and humankind should adopt a protective attitude towards other forms of life.

Deep concern has been expressed at the tendency to usurp more and more of the resources of the planet for human beings leaving less and less for other forms of life. According to Foster writing in *The Vulnerable Planet,* 'Human beings now use (take or transform) an estimated 25 per cent of the net photosynthetic product (NPP) i.e. the plant mass fixed by photosynthesis—over the entire earth (land and sea), and 40 per cent on land'. As human beings take more of the primary productivity of the earth for themselves, less is left over for other species. According to Meadows and Randers, the authors of a path breaking study titled *Beyond the Limits,* 'Somewhere along the path of NPP usurpation, there lie limits. Long before the ultimate limits are reached, the human race becomes economically, scientifically, aesthetically, and morally impoverished'.

Relationship of Present Generation and the Next

Every generation inherits a certain world from its ancestors, and leaves behind a certain world for the next generation. A criteria for evaluating any generation is whether it leaves behind a better world than the one it inherited. If we examine this question carefully and honestly, then the present generation cannot escape the verdict of history that at no other time has any generation left a more dangerous world for the next generation. For the present generation, there may be a unique indictment not faced by any previous generation that this generation allowed life-creating conditions to be disturbed, perhaps permanently, in the form of the worsening threat of climate change. In addition, there is the worsening threat from WMDs, terrorism, the danger of WMDs falling into the hands of terrorists, the rapid spread of several hazardous technologies and substances etc.

In social life also, the next generation will suffer the adverse impacts of growing disintegration and alienation. The available information on a wide range of indicators of social disintegration tells us clearly that this is increasing very rapidly. Perhaps more rapidly that ever before. Many of the scars are likely to haunt us long into the future, as the victims belong to the young generation.

Richard Douthwaite has summarised the results of some important studies in UK, 'Divorce has also had long lasting and extremely damaging consequences for the children affected by it. The National Survey of Health and Development, which has followed the fortunes of 5,362 babies born in the same week in March 1946, assessed children from broken homes in their early twenties and found that half were underachievers, educationally disadvantaged and in poor jobs, a much higher proportion than was the case for children whose parents' marriages were intact. The children of divorced parents also drank and smoked more. They lacked self-confidence, put themselves down, and were often still very angry. Delinquency was much higher, particularly among those whose parents had spilt up before the children were five.

In its 1995 report on the state of the world's children, UNICEF drew up two visions for the year 2050. In vision 1 there is increasing poverty, disease, disaster, pollution and violence, while in vision 2 there is a significant reduction of all these problems. After presenting both these visions, the authors of this report said, 'The choice between these futures must be made not in 50 years time but today. More than two decades when the world could have been addressing these urgent problems have already been lost. Another lost decade will probably be decisive'.

PITFALLS OF PROGRESS

When time-honoured practices of maintaining soil fertility are neglected, soil fertility is impeded. The solution lies in correcting these mistakes, not in applying chemical fertilisers, for chemicals disturb and disrupt the natural process in numerous ways, for eg chemical farming kills earthworms, whom Charles Darwin called 'the builders of civilisations' and who play an invaluable role in protecting and maintaining the fertility of land.

Something to Think About

The Myth and Reality of Progress

TILL JUST a few decades back, the concept of ever-continuing progress had a strong hold among reputed scholars. In his famous lectures published under the title *What is History*, Prof EH Carr has given two quotes to show how well entrenched this view was. The first quote from the introduction to the first volume of *The Cambridge Modern History*, said, 'We are bound to assume, as the scientific hypothesis on which history is to be written, a progress in human affairs'. The second quote was from the last volume of this history, in which historian Dampier felt no doubt that 'future ages will see no limit to the growth of man's power over the resources of nature and of his intelligent use of them for the welfare of his race'. Today, with the experience of recent decades behind us, this view is likely to be hotly contested. While most people would agree that 'man's power over the resources of nature' has indeed been increasing rapidly, they are also likely to raise huge questions about whether man has made 'intelligent use' of this power for the welfare of his race (not to mention other forms of life).

While Carr himself also was optimistic about the future of humanity, he nevertheless raised some important questions. He said:

> At the present time, few people would I think, question the fact of progress in the accumulation both of material resources and of scientific knowledge, of mastery over the environment in the technological sense. What is questioned is whether there has been in the twentieth century any progress in our ordering of society, in our mastery of the social environment, national or international, whether indeed there has not been a marked regression. Has not the evolution of man as a social being lagged fatally behind the progress of technology?

This as well as even more disturbing questions have been raised about the 20th century which despite all the amazing advances of science and technology also turned out to be the most violent and destructive century.

Chris Brazier, who has written a short history of the world, has summarised a lot of data to tell us that the scale of devastation of human life in the last hundred years should alone be enough to dispel any lingering illusions about the inevitability of progress. During the 20th century some 150 million people died in war. About 100 million died in the great famines of the century. A further 100 million died as a result of government repression.

It is even more important to stress that these statistics can never come even near to capturing the pain and suffering of the victims of the most cruel weapons that humankind invented in this century, or the lasting distress suffered by those who lost their near and dear ones. All this, after a point, is truly beyond description.

The philosopher Isaiah Berlin said of the 20th century, 'I remember it only as the most terrible century in human history'.

If we go back farther to about 500 years or so, when in the post-Columbus world the foundations of the modern world were being laid, we again see a series of unforgivable but also unrepentant crimes against humanity unfolding one after another.

At the time of Columbus' first voyage in 1492, the Americas were the home of nearly 100 million people (compared to a European population of only about 70 million at that time). Within the next 100 to 150 years, nearly 90 million of these indigenous people perished. Judging by the world population at that time (about 600 million) this means that nearly one seventh of the world population was finished off within 100 to 150 years. Many of them died in wars started or instigated by the newcomers and many others died in the epidemics brought by them.

Columbus forced the Taino Indians in Hispaniola to bring him an ounce of gold every three months. Those who did not, had their hands chopped off while escapees were hunted down with dogs. Regarding the impact of the advent of the 'civilised' people in the new world, the priest Bartolome de Las Casas wrote: 'For 40 years, they have done nothing but torture, murder, harass, afflict, torment and destroy them with extraordinary, incredible, innovative and previously unheard of cruelty'. Las Casas estimated that 50 million Indians perished in Latin America and the Caribbean within 50 years of Columbus' landing.

According to Wayne Ellwood who prepared a special issue of the *New Internationalist* on this subject:

> Scholars now reckon that 90 per cent of the indigenous population of the Americas was wiped out in a century-and-a-half—the greatest demographic collapse in the history of the planet and the proportional equivalent of nearly half a billion people today.

The genocide was later repeated in Australia. According to Robert Huges, author of the *Fatal Shore*:

> It took less than 75 years of white settlement to wipe out most of the people who had occupied Tasmania for some 20 thousand years.

However, soon there were so many deaths among native Indians that slaves had to be obtained from Africa to toil in the new world. Wayne Ellwood estimates : The Atlantic slave trade lasted nearly 400 years during which time 15 million Africans were shipped to the Americas. Conditions during the crossing were so horrific that between half and a third of the African captives died en route. Of the rest, most died within a decade of their arrival in the strange new land.

This was followed in the 19th century by a rapid phase of annexing African territories, which included brutal annihilation. In a period of only 20 years, the population of French Congo was reduced by one-third of its former size. When the Germans established themselves in South West Africa, more than half of the population was exterminated. In Asia perhaps, the world's largest ever plunder was accompanied by equally ruthless violence and deceit. Finally, the fierce competition for colonies within the imperialist powers led to the first world war.

As this short historical review makes it amply clear, in

any review of progress, it needs to be asked who prospered and who were sacrificed at the altar of prosperity of others. Such questions remain highly relevant in the 21st century as inequalities have increased further in many countries.

It is only by raising questions about people and groups who had to suffer so much that any notions of 'progress' can be evaluated in a just way. Perhaps, what is even more essential is to study not only the condition of all human beings at various time periods but to instead look at all life forms. If instead of asking how the life of human beings has changed during the last five hundred years, we ask 'how have all species changed during the last five hundred years', then we will get a very negative answer. The reason is that the natural habitats of all life forms including forests, oceans, rivers, estuaries and coastal areas, wetlands etc. have been damaged on a massive scale largely by man's activities. In addition, various species have also been hunted or overfished to the point of extinction or high levels of depletion.

Indeed, the biggest question marks on the inevitability of progress arise due to environmental problems particularly climate change.

The Global Environment Outlook (GEO) Report prepared by the United Nations Environment Programme has said on a note of warning, 'In the future, the continued degradation of natural resources, shortcomings in environmental responses, and renewable resource constraints may increasingly lead to food insecurity and conflict situations. Changes in global biogeochemical cycles and the complex interactions between environmental problems such as climate change, ozone depletion and acidification may have impacts that will confront local, regional and global communities with situations they are unprepared for'.

Commenting on the current development pattern, the GEO report declares quite explicitly: 'The use of land, forest, fresh water, coastal areas, fisheries and urban air is beyond their natural regeneration capacity and therefore it is unsustainable'. Further this report says, 'Global developments in the energy sector are unsustainable'.

The World Resources Report (WRR), prepared by the World Resources Institute with the UNDP and UNEP said, 'The world is not now headed toward a sustainable future, but rather toward a variety of potential human and environmental disasters'.

The 21st century may witness the pinnacle of technological advances yet it cannot be denied that at the same time the prospects of sustainable progress are more threatened today than ever before. The sooner we recognise this, the better we will be prepared to meet the tough challenges ahead.

How Scientific is this Progress

Destructive Development Ignores Scientific Reasoning

SCIENTIFIC PROGRESS implies gaining a better understanding of nature to ensure that the needs of human beings and all other forms of life are met on a sustainable basis, without disturbing too much (and certainly not destroying) the life-nurturing conditions of nature and her many kinds of habitats. This is achieved by improving the understanding of how the various aspects of nature exist, work and interact with each other.

If the path of scientific progress is defined in this way then it is very clear that presently humankind is not following this path. In recent years, human beings have ruthlessly destroyed nature, her many habitats and the life-nurturing conditions.

There is need for wider acceptance of this failure so that grave mistakes of the recent past can be admitted and we can embark on true scientific progress (as defined above).

For example,the foremost step in a scientific study of a river system is to form comprehensive understanding of what important environmental role the river and its tributaries perform in maintaining the balance of nature, and how fish and other forms of life flourish in these rivers and their immediate surroundings. This should of course include an understanding of how various human settlements around rivers obtain various life and livelihood benefits from rivers.

On the basis of this understanding, scientific progress consists in ensuring that the life and livelihood of human beings continue to benefit from river flows on a sustainable basis, without harming the natural and environmental functions of the river systems and without threatening the fish and other various forms of life flourishing in and near rivers. All issues relating to river systems—whether sand mining, dams and barrages, irrigation, hydel power, flood-control navigation etc. have to be taken up on this understanding, only then we can say that scientific progress is taking place. But what is taking place is not at all scientific in this sense. Although highly qualified engineers and geologists may be involved, what is essentially taking place in most places is a reckless violation of scientific study and progress. Quite often, big business interests take decisions which are most lucrative for them, and then somehow a cover is provided for this with cooked up 'scientific' studies.

Such is the experience of not only river systems but also most other areas and contexts. In the case of agriculture, perhaps the most important aspect in a scientific study of sustainable agriculture is to understand how nature works to maintain soil fertility. This is described here in the words of Sailendra Ghosh, a writer who repeatedly emphasised the need for organic agriculture. He writes, 'Let's first take a look at the bounties of nature in an undisrupted system. Firstly, nature has made ample provisions for the supply of nitrogen through the root system in a variety of ways. Rains wash down the nitrogen generated by thunder. Bacteria in the nodules of leguminous plants capture nitrogen from the atmosphere and fix it into the soil. So do some ferns, some forms of algae and some genera of free living bacteria. Various genera have been endowed with this quality so that this operation could continue under differing conditions. Probably the most important source of supply of nitrogen as also of other nutrients is the decomposition of animal and plant wastes. This is done by decomposer bacteria and fungi'.

Animals eliminate excess organic phosphorus by excreting phosphorous salts in urine. There are also a phosphatising bacteria, Sailendra Ghosh says, to convert phosphorus into stable forms of phosphate salts which remain bound with soil. At the beginning of rains, the mineralisation of organic matter releases phosphates for uptake by the plants.

Further, Ghosh writes, 'For the supply of all other macro and micronutrients as also of vitamin and plant growth promoting substances, there are elaborate arrangements in natural soil systems. The soil abounds in countless forms of micro organisms—bacteria, fungi, viruses, protozoa, yeasts, algae etc. to perform different but interrelated functions. Each produces different kinds of enzymes by which all stand to benefit'.

To enable this beautiful but complex natural system to carry out its work of maintaining land fertility, we should protect forest cover in and around agricultural fields, we should return organic waste to soil and we should choose carefully mixed cropping and rotation systems.

When all these time-honoured practices of maintaining soil fertility are neglected, soil fertility is impeded. The solution lies in correcting these mistakes, not in applying chemical fertilisers, for chemicals disturb and disrupt the natural process in numerous ways, for eg chemical farming kills earthworms, whom Charles Darwin called 'the builders of civilisations' and who play an invaluable role in protecting and maintaining the fertility of land.

However, organic farming should not be equated just with 'non-chemicalisation' as, in the words of Sailendra Ghosh, 'Organic farming means farming in the spirit of organic relationship. When you say this, it opens up a whole new vista. In nature, organic relationship is a pervasive phenomenon. Everything is connected with everything else. Since organic farming means basing farming on integral relationships, we have to know the relationship between the soil, water and plants, between soil, soil microbes and waste products, between the vegetable kingdom and the animal kingdom (of which the apex animal is the human being), between agriculture and forestry, between soil, water and atmosphere etc. It is the totality of these relationships that is the bedrock of organic farming'.

This is the scientific approach, but motivated interests intervene in policy making so that only those policies which favour chemical fertilisers are promoted.

A leading economist of India, Ashok Rudra has noted,

> The policy of the Government of India with respect to soil nutrients since the beginning of the sixties has been one of vast and continuous expansion of the use of chemical fertilisers by all possible means. All the policy instruments in the hands of the government, such as import control, industrial licensing policy, fiscal policy, price control

> policy etc. have been applied in a concerted manner so as to achieve that aim.

Further he said, 'There has, however, been no noticeable effort whatsoever for the maximum mobilisation of organic manurial resources'.

Even the government's own economic survey admitted, 'The easy availability of chemical fertilisers has tended to reduce the attention which farmers, agricultural scientists and administrators must continue to devote to organic manures and plant materials'.

The green revolution phase has generally involved the spread of intensive monocultures, but this is harmful for the long term fertility of land. As the World Resources Report (WRR) says: Soils under intensive monoculture tend to lose organic matter, their ability to retain moisture thus becoming more susceptible to erosion and ultimately losing of their fertility and productivity'.

Spread of intensive monoculture generally involves a higher reliance on chemical pesticides. A very small part of the pesticide applied on a field—less than 0.1 per cent in many insecticides—actually reaches its target organism. The rest plays the role of polluting the land and water poisoning birds and other forms of life. As WRR says, 'wholesale elimination of helpful soil dwelling insects and micro organisms that build soil and plant nutrition sometimes occurs, essentially sterilising the soil'.

The World Resources Report has stated clearly that in the 50 years since pesticide use became widespread, the percentage of crop loss from pest damage has not measurably declined. Insects, weeds and plant diseases still claim 30 to 35 per cent of total crop production today—about the same percentage estimated for the pre-chemical age. Today, more than 500 insect and mite species are immune to one or more insecticides, 113 weeds are resistant to one or more herbicides and 150 plant pathogens are resistant to fungicides.

A *FirstPost* report on Punjab says: The Punjab soil and water conservation department says that with the wheat and paddy cropping pattern introduced by the Green Revolution, soil quality has degraded and water resources depleted. 'If this trend continues, these resources are likely to degrade and deplete even further, and we will not be able to sustain the production levels we have achieved thus far', its report read. 'The soil in Punjab has been degraded by its over-exploitation to maximise food production. Excessive use of chemical fertilisers, insecticides, pesticides, etc. has destroyed the physical structure of the soil, leading to its decreased water holding capacity and friable and loose structure'.

It further noted, 'The macro and micro fauna and flora of the soil have been affected, and soil building agents have been largely affected, leading to poor fertility and production capacity of the soil. This has also led to emerging biological (insects, diseases, weeds) and physical (water, soil, salinity etc.) stress, thereby increasing costs of input and declining productivity and profitability levels'.

It adds, '(With) the opening up of the globally competitive and quality-conscious market, it is doubtful that the farmers of Punjab would be able to survive in this competition by continuing with the rice-wheat based economy'. The Green Revolution was also steered by American policy advisors to the then government in India, following the PL 480 aid to India.

Such evidence reveals how in most crucial issues relating to agriculture the scientific approach has been abandoned to yield place to highly unscientific approach under the impact of motivated interests.

Such examples can no doubt be multiplied, but the essence of the argument is to emphasise that in most policy related issues impacting millions of people, the scientific approach is being sacrificed on the altar of commerce and greed.

Without Plundering Others

Alternative Understanding of Economic Development

IDEALLY ANY economy should try to achieve the following:

- The basic needs of all its members should be met.
- All the adult members should be employed creatively in those jobs and activities that are fairly close to their chosen interests and skills to meet these basic needs.

It is the role of economists, managers and planners to meet these two objectives in such a way that the creative work of the people is linked to meeting their basic needs (and a few small comforts).

It is important to emphasise that this should be achieved without plundering any other economy or nation. The fulfilment of needs and a few small comforts should never be at the cost of depriving other societies from a similar aim.

This is why we emphasise that the objective is only to

meet the basic needs and a few small comforts. We never say—provide as many comforts and luxuries as possible to all people. If an economy seeks to provide great comforts and luxuries to all its people, then in all probability this will be achieved by plundering the resources of other economies or nations, depriving their people of basic needs. In addition, the resources within the economy will be recklessly used so that the needs of future generations will be sacrificed by the resulting ecological ruin. This should be clearly opposed and rejected.

Therefore it is clear that the economy should strive to meet only the basic needs and small comforts of its people. We do not intend to be unnecessarily restrictive. This is why we add 'small comforts' to basic needs. So the joy of dolls and toys, birthday parties and picnics, honeymoons and bridal dresses are not ruled out in our ideal economy—our only plea being that all this should be within reasonable limits.

The market forces will be given adequate role, but this will be counter balanced by strong people's movements which support environment protection and oppose the exploitation of other societies. There should be a strong movement which tells people how the interests of next generations—our own children and grandchildren—are best served by a simple life which does not impose a heavy burden on natural resources. In addition, there should be strong movements to protect consumer rights and labour rights.

In rural communities and small towns, an attempt should be made to meet most of the basic needs on a self-reliant basis by organic farming, artisans, cottage scale industry and local small scale industry. All the remaining goods can be purchased from the wider market. Big private companies will also play an important role in the production of those goods for which they are best suited but no one will be allowed monopolistic control, or unfair practices to gain dominance. The public sector will play an important balancing role in the creation of infrastructure and a few important consumer goods like essential medicines. Unnecessary and bulky spread of government in non-essential economic activities will be avoided.

Both private and public corporate sectors will have to adopt transparency and public accountability in their functioning. There will be strict checks and balances to prevent corruption and misuse of power, including monopolistic tendencies.

Saying Yes to Need, No to Greed

The concept of 'basic needs' is one of the most widely used concepts and yet this is also one of the most elusive concepts. Everyone seems to agree that 'basic needs for all' is a highly desirable objective, but at a practical level there are several difficulties as people have very wide ranging views on what constitutes basic needs.

The concept of basic needs has a very bloated and exaggerated meaning for an increasingly large number of people. Goods and services without which they cannot live (or think they cannot live) are so many that their 'needs' include what millions of poor people would regard as the biggest luxuries. Why does the modern man's perception of a satisfactory life include such a wide range of goods and services, a range that appears to be increasing all the time? Is this because we've lost something precious and try to fill this gap in our life by increasing acquisition of goods and gadgets? A desperate, but also futile, effort to fill a social and spiritual void with material goods?

It is amazing how everyone appears to be in such a big rush, and how little actually gets done to improve life. Several of us manage to convince ourselves that our very busy work schedule is of great significance, and to support this we hasten to acquire as many goods and services as possible which we perceive as necessary for this lifestyle.

A very strong force that leads to a bloated sense of needs is the constant advertising that bombards us from all side at all times. Advertising does have a role as far as giving relevant information to people about useful products is concerned, but present day advertising goes much beyond this limited role and constantly makes us yearn and crave for goods we do not really need. Advertisers take advantage of every conceivable human weakness and play on human emotions in the most reckless way to sell all sorts of products of dubious merit, or even clearly harmful products. Billions of dollars are spent every year in the advertising of liquor, cigarettes, junk food and useless or even hazardous drugs, cosmetics and pesticides.

Partly as a result of our own susceptibilities and partly as a result of heavy advertising, we've been steadily commoditising our life, vainly trying to fill social and spiritual voids with material goods. Understanding this is the first step towards changing our lifestyle in such a way that our needs are not in contradiction with the social objectives of environmental protection and meeting the basic needs of ALL people.

Suitable changes in government policies and public utilities can also contribute considerably to reducing excessive private spending. Health and education are the most obvious examples. Availability of good public transport can reduce the need for privately owned vehicles. Community gatherings and celebrations can reduce the need for privately organised expensive entertainments.

Thus a combination of changes in personal lifestyle,

enhanced role of community life and relevant changes in government policy can help us to modify our understanding of basic needs in tune with desirable social norms. The present day society and the state of environment simply cannot afford a very bloated view of needs. It will be difficult to fulfill an exaggerated sense of needs for everyone without inflicting heavy damage on already badly mauled environment. So reaching a realistic understanding of needs is essential if we want to create a world where basic needs of ALL can be met on a sustained basis without damaging the environment too much and protecting the interests of future generations as well.

The idea that the production of material goods and services is desirable only up to a point beyond which the pursuit of this objective should be given up voluntarily may appear strange to some growth addicts but it is deeply embedded in teaching of several spiritual leaders and social reformers who had very large followings. If we can get rid of the dominance of some narrow-minded economic thinking of relatively recent origin and give due importance to the great spiritual and social movements of human history, then there are many sources of support for this idea. By drawing on these inherent strengths of various cultures and societies, we can build a new world which is free of greed and hence of most type of violence, where the basic needs of all people are met, where human relationships are valued, where there is genuine regard for other forms of life and for protecting the natural environment for this and future generations.

Equality and simplicity are two concepts which acquire their real significance only when taken together with each other, but together they help in building a better world.

Thus equality and simplicity (self-restraint) are the two pillars on which the foundation of sustainable development can be built. To be able to achieve this, very powerful vested interests have to be confronted and overcome, and very deep rooted economic and political changes both at global level and at the local level have to take place. A well-thought out, dedicated and sustained campaign to convince more and more people that both happiness and security are best achieved by pursuing sustainable development based on equality and simplicity is the biggest need of our times. Once most people are convinced of this, even the most powerful vested interests will not be able to stop the path of sustainable development.

The concept of development needs to be freed from the terrible misconception of more and more accumulation and consumption. As long as production of material goods is increasing in a society to meet the basic needs of all, this is justified and in conformity with the welfare of this society. But significant production and consumption of material goods too much beyond this cannot be justified and should be called 'wasteful destruction', not development.

Development should be redefined as the collective endeavour to reduce distress and suffering of all forms of life as much as possible. Some distress and suffering will always be there, but most of the distress and suffering of this world is caused by avoidable causes. Development needs to be seen as the efforts of human beings to check these avoidable causes of distress and thereby reduce the distress of all forms of life as much as possible.

This endeavour should take good care of the needs of future generations. We should protect the resources base so that future generations of human beings and other forms of life can meet their basic needs equally well or in even better ways. This is the essence of sustainable development.

New Ethical Economics

Tapping Creativity To Meet Real Needs

THE MOST significant concern of any economy should be to meet the basic needs of all people. The real challenge is to tap the creativity of people in such a way as to link it with meeting the basic needs of all. The economy should be geared to providing basic needs (and a few small comforts which bring joy and colour to human life) on a priority basis. The use of luxuries should be discouraged by taxing them heavily.

There are obvious limits as to how many industrial and other goods can be produced for human beings without damaging environment seriously and denying the needs of other forms of life. If the production of luxuries is given free play, then this may be at the expense of meeting the basic needs of many people. And if there is too much production for human beings (much beyond basic needs) then this may be at the expense of meeting the needs of other forms of life. As this destroys environment, the capacity of future generations, including our children and grand children, to meet their needs may be badly impaired.

People want to do creative work, work which meets real needs and gives them satisfaction. This creativity of people needs to be linked to the production of goods and services which meet the basic needs and small joys of people. Trying to establish such a link is the real job of economists and not the mostly stupid, wasteful things they do presently.

The habitats of many forms of life have been ravaged badly in recent decades. Human beings caused this horrible damage but they alone are capable of repairing it. So the creativity of people should also be used with a sense of urgency for cleaning lakes, rivers and seas; for protecting forests, planting more trees, saving wetlands and pasture-lands and in other ways.

It is important to keep in mind the needs of future generations as well. In recent decades, the fertility of a lot of good farmland has been badly impeded by the indiscriminate use of industrial chemicals. The threats of many kinds of species extinction and climate change are even greater. So it is important also to link the creativity of people with protecting the environment in such a way that the interests of future generations are safeguarded.

The economy should link the creativity of people with meeting the basic needs of people, protecting habitats of other forms of life and healing nature in such ways that the life of future generations is protected. In other words, people should find satisfactory livelihood in such tasks of priority as meeting the real needs of people, protecting environment and other forms of life. While market-mechanisms should continue to play an important role in economy, this should be only within a framework provided by the government which emphasises basic needs for all and discourages highly wasteful consumption and accumulation by a few. The government should not withdraw from the responsibility of ensuring the basic needs of all, and providing conducive conditions in which the creativity of people can achieve this aim.

Rural communities should be encouraged to become self-reliant in meeting their basic needs on a sustainable basis. They should have the right to decide how natural resources in and around the village are used on an egalitarian and sustainable basis.

International trade should have a limited, relevant role to enrich the life of people, but under no condition should it be allowed to disrupt the livelihoods of entire communities.

Small and medium scale private entrepreneurship should play an important role in the economy. The corporate sector will have to operate within clear requirements of responsible behaviour towards labour, environment and consumer rights. The growth, particularly monopoly growth of corporations beyond certain levels should be discouraged. The public sector should play an important role in meeting some important needs of people and to act as a check to irresponsible behaviour by sections of the private sector (for example private sector companies selling medicines at an irrationally high price).

Human beings alone have the capacity to work in a planned way for the protection of all forms of life. Their basic role on earth is to protect and promote the welfare of all forms of life (including, of course, fellow human beings). The economy should function, keeping in view this basic principles so that people can find satisfactory livelihood in meeting the basic needs of all people, protecting environment and other forms of life.

Food and Farming System

To Feed the Hungry and Protect the Environment

IT IS widely agreed that increasing land availability to the poorest peasants and reducing inequalities in the distribution of agricultural land are highly desirable for reducing hunger and malnutrition in the world. Despite this, inequalities in the distribution of agricultural land remain at high levels in many countries probably reaching the peak in some Latin American countries.

Inequality is often expressed by a statistical measure called the 'Gini Coefficient', which varies from zero (equal assets for everybody) to 1 (one person owns everything). In most Latin American countries, the Gini Coefficient for land distribution is around 0.8. The inequalities in agricultural land may be most acute in Latin America, but these also exist to a significant extent in most other countries of the world.

In a widely quoted publication titled 'Agriculture Towards 2000' the Food and Agriculture Organisation has emphasised that more equal land distribution is likely to increase productivity of land,

> It is important to stress here that yields per hectare are as high on small as on large farms or, under traditional agriculture, even higher. With a few notable exceptions, total output per hectare is higher on small farms, chiefly because their intensity of land use is higher. A more equal distribution of production inputs, including services, can only help to strengthen the role of the small farm in expanding production. The fact that some attempts at radical redistribution of land have led initially to lower production does not invalidate the conclusion that after some years a more even distribution of farming resources and inputs should help rather than hinder growth of output.

This view of the FAO is supported by a six-country study by the International Labour Organization (ILO) which estimated that 'If land were equally distributed among all agricultural families (including the landless), and the new equal holdings achieved yields equal to present holdings of the same size and used a similar level of inputs, food output could potentially rise by anything from 10% (Pakistan) and 28% (Colombia and a rice-growing Malaysian region) to 80% in northeastern Brazil. Such a radical redistribution is, of course, rarely attempted, but the figures indicate the theoretical potential'.

Therefore, the evidence in favour of small farms and for having a more equal distribution of agricultural land is quite strong. It is another matter that as vested interests opposing redistribution are quite strong, the land reform effort can face several obstacles due to which there can be temporary setbacks. Such resistance can come from large landowners as well as corporate interests promoting large-scale farming of commercial crops with a big share of export crops.

The distortions created by local inequalities and an unequal international trade system in fact feed on each other. The local powerful elites are eager to get luxury consumer goods while foreign companies are only too happy to meet these demands while gaining access to the cheap land and labour of these countries. The ultimate victims are the marginal groups on whom all the pressure of meeting the unreasonable demands of local and foreign elites is passed on. So empowerment of these poor and marginalised groups in their struggles to save their livelihood is the most important factor in fighting hunger and famine. They do not need emergency food aid shipments (which in any case will probably never reach them in time) as much as the support of solidarity groups to save their livelihood base from the onslaughts of selfish, powerful interests at home and abroad.

In Africa, in the wake of the growing concern over diminishing per capita production of food, some ambitious food production schemes were initiated with the support of international aid agencies but they failed to meet the needs of the worst affected, most needy, precariously placed groups. As Barbara Dinham and Colin Hines wrote in a critique of these schemes in the *Ecologist,*

> Large-scale food producing by-passes the problems confronting peasant communities who have been moved into smaller and less fertile land, who are not paid a sufficient price for the crops they produce for the market, who are ill-served by distribution of storage networks, and whose needs for investment in education, health and water supplies are ill-met.

The real concern should not be just to increase production but to increase it on such farms, in such ways as to meet the food needs of the most needy households. Otherwise, we will face such cruel distortions as those seen at the time of the Sahelian famine when vegetables were being airlifted for exports from famine affected countries! The need is not just for more food but for food which reaches the hungry of the world.

In recent years, as growing concern has been expressed about the hazards of excessive use of agri-chemicals particularly chemical pesticides, there has been increasing interest in alternative technologies which reduce or even eliminate the use of agri-chemicals while relying more and more on various natural processes and organic materials to get good yields, thereby also maintaining long term fertility of land.

According to the World Resources Report, 'When on-farm and off-farm soil and surface water resource cost were included, resource-conserving farming out-performed conventional approaches by almost a two-to-one margin in net economic value per hectare (including off-site environmental costs)'.

In Indonesia, a very instructive effort was launched in 1986 to avoid excessive chemical sprays of rice crops and replace the heavy dependence on chemical pesticides with a many-sided effort called 'integrated pest management' which included many non-chemical methods of facing the threat of pests. As Peter Kenmore, a USA scientist closely involved in this effort explained less than 2 years after launching this effort, the use of chemical pesticides declined steeply while the yields of rice increased to a significant extent.

According to case studies of successful vegetable and rice farms using ecological methods in Philippines, in the largest set of adjacent farms totalling 1,000 hectares using the biodynamic farming method there was a yield increase of 50 to 100 per cent and an increase in net income by farmers of 200 to 270 per cent, compared to the green revolution methods. Nicanor Perlas, a Fillipino agricultural scientist, has said while presenting these case studies that *a rapid transition from chemical farming to sustainable agriculture is possible if correct technical principles are followed.*

Growing a wide variety of indigenous crops should be encouraged. The richest knowledge of the biological diversity of a region is likely to be available only with local farmers. Eminent Indian rice scientist, RH Richaria, identified nearly 17,000 rice cultivars and varieties in Chhattisgarh region of India with the help of farmers, particularly tribal farmers, and his co-workers. About the wisdom of the farmers he wrote,

> Invariably I found in rice areas some rice growers taking keen interest in their local rice varieties as they are very much absorbed in them and they have all praise for them, so much so that they trace back the history of individual rice varieties to their ancestry with their utility... I also observed that some of them identify their rice varieties in their own way (not in terms of the modern knowledge of Botany) which amounts to thousands. This inherent and intuitional facility of selections and maintenance of thousands of rice cultivars gradually being accumulated and descended upon for unknown centuries, ever since rice first originated must be preserved... Some of these varieties of rice were known for their high yield, some for their great cooking qualities, some for their aroma and some for other cherished qualities.

In particular, an effort should be made to support those locally useful crops and crop varieties which have been displaced thoughtlessly in recent years. Legume crops deserve special attention. Frances Moore Lappe and Joseph Collins, write in *Food First – beyond the myth of scarcity*, 'the beans and corn diet of Latin America, the lentils and rice of India, and soyabean and rice diet of China appear to most Americans as starchy and nutrient poor. In fact they are not. Such diets evolved because they work. As basic dietetic staples, these combinations are, in fact, quite ingenious. In each case the two items together give more biologically usable protein than if each were eaten separately'. In view of this importance of mixing cereals and legumes to get a balanced diet, some experts have gone to the extent of referring to the diminishing production of pulses (or other legumes) as the divorce of agriculture from nutrition. This is particularly bad for the weaker sections who don't have access to other, more expensive proteins. Therefore, top priority should be given to increasing the production and availability of pulses.

An effort should be made to look at the entire agro-eco system instead of examining individual crops and grain yield in isolation. As an organic farmer of India, GR Iyengar has said,'Few of us realise the havoc that modern agricultural practices are wreaking on our countryside. Farmers have forgotten the habits that supported a wide variety of wildlife and countless varieties of wild plants, flowers and trees that are essential for profitable and sustainable agriculture. What is happening in farming today is that the agriculture technology is acting in isolation, treating organisms in isolation, which leads to a disturbance of the natural system of checks and balances. It should be a sensitive balance of organisms in nature that should be allowed to spread. Few people realise today that there is a certain symbiosis between the various elements of nature like between flowers and pollen, soil and organisms. The role of the ecological balance in managing habitat has to grow'.

Reversing degradation of land which has been continuing for so long will involve several bold initiatives, including some which can be expected to evoke a lot of resistance among strong vested interests. Yet there is no doubt that such initiatives have to be taken and cannot be delayed for too long. Protection of our precious cropland and soil is too important a task to be neglected any longer, particularly keeping in view the needs of the next few generations.

As a scientist BZ Rozanon has said so movingly:

> The task of agriculture is thus not confined to obtaining the biological product but extends to constant maintenance and augmentation of soil fertility. Otherwise, we will very quickly consume what by right belongs to our children, grandchildren, and great-grandchildren, to say nothing of more distant descendants. It is this—that our generation lives to a certain extent at the expense of the coming generation, thoughtlessly drawing on the basic reserves of soil fertility accumulated in the millennia of the biospheric development, instead of living off the current annual increment—that causes the increasing concern of scientists dealing with the state of the planetary soil cover.

The World Commission of Environment and Development had recommended, 'The legislative, policy and research capacity for advancing non-chemical and less-chemical strategies must be established and sustained'. However, in practice, few countries have followed this advice. In principle it is widely agreed that agricultural polices should be guided by the objectives of protection of environment and sustainability. But in the real world, are agricultural policies really guided by these noble objectives? There is a growing suspicion that narrow-minded, short-term profit considerations have a far more powerful influence in real life farm policy decisions.

The World Resources Report minces no words in making a firm statement on this issue: 'Current farm practices in industrialized countries have created incentives for farmers to use environmentally damaging practices and in many cases, penalized farmers for switching to more sustainable practices'. For example, the report says, the system of farm programme payments in the USA has worked against long-term rotations and reduction of chemical inputs. In the European Community, price policies have discouraged production of pulses. Although mixed crop-livestock systems can be the basis for environmentally sustainable farming, distorted price structures have tended to push agriculture in the other direction.

Over 20 UK farming, consumer, organic, animal welfare, environmental and Third World groups have formed the Sustainable Agriculture, Food and Environment (SAFE) alliance. According to SAFE's campaign statement:

> Agriculture is about more than simply producing food. It is a way of life and makes a vital contribution to the health of rural communities. SAFE seeks to switch farm subsidies away from price support towards payments for sustainable and environmentally enhancing farm management practices agreed on a whole farm basis. All the land on any one farm would be included in the scheme, and payments made would be tiered on an acreage basis.

The effect of these whole farm management agreements, argues SAFE, would be to put smaller family farms (the mainstay of many rural communities) back on a level playing-field with much larger farms, and to remove the present in-built bias towards increased farm size. Such agreements, argues SAFE, would both encourage participating farmers to modify their production methods to take full account of environmental factors, and also reward those, such as organic farmers, already practising environmentally-sensitive methods. The SAFE Alliance has tried to define the requirements of good agricultural system:

- An agriculture that is supportive of rural communities, that halts the decline in full-time farm employment and provides a stable livelihood for farmers and farm workers;
- An agriculture that does not jeopardize the health of those who work or live on the land or the consumer through the use of polluting or toxic production methods;
- An agriculture that is capable of flexible response to national food and nutrition goals designed to improve public health;
- An agriculture that produces affordable food of high nutritional quality and that minimizes chemical and microbiological contaminants;
- An agriculture that does not lead to the reduction of soil fertility, that minimizes reliance on non-renewable resources and that is sustainable;
- An agriculture that both conserves and enhances the countryside, not only in its visual aspect but also in terms of its resources and wildlife;
- An agriculture that respects the welfare needs of farm animals;
- An agriculture that does not threaten the development and maintenance of food security and sustainable agriculture in other countries, especially those in the Third World.

A good food and agricultural system is one which makes available satisfactory livelihood to all members of the farming community and wholesome, nutritious food to all people in a sustainable way ensuring welfare of farm animals, protecting the soil and its fertility for future generations. Unfortunately, at present, there is a lot of economic tension and uncertainty among farmers not only in developing countries but also in some of the richest countries like the USA. A significant share of the farming community is landless (or near landless) and it is deprived of a fair share of the farm income. Many landholders are themselves in danger of losing control over their land.

Again even in the richest countries, care of soil is badly neglected. The machines that have taken over most of the work on farms can produce more food but cannot protect soil for future generations. As more and more skilled peasants are being driven away from agriculture, it becomes more and more unlikely that enough people will be available not just to produce food but also to protect land. Farm animals are being treated just as a commodity with least concern for their welfare.

On the consumption side, not only do poverty and inequalities deny adequate food to a large number of people (probably in the neighbourhood of one billion) but even those with enough purchasing power find it increasingly difficult to get wholesome, nutritious food due to the domination in markets of food which has been heavily treated by chemicals or subjected to other processing (or adulteration) which have peeled off valuable nutrients while adding some harmful substances.

So the basic priorities of improving the farm and agriculture system should be:

- Empower those who work on farmland to maintain their close links with land.
- Support them to get an adequate livelihood from cultivating the land and protecting its soil.

- Establish harmony between the welfare of farm animals and economic returns of farming. In other words, make such productive use of farm animals as can bring economic returns without causing any cruelty to them.

- Establish direct links of farmers and consumers so that consumers can get good quality food and farmers can get a fair price in return. Monitor trade, processing and middlemen interests carefully to ensure that interests of farmers and consumers are not violated.

Health Hazards of Meat

A study by Dr Neal D Barnard, Dr Andrew Nicholson and Jo Lil Howard in the USA has linked regular consumption of red meat and poultry, in particular, to significant increases in the risks of developing high blood pressure, heart disease, diabetes, gallbladder disease, overweight and resulting osteoarthritis, food poisoning and cancers of the colon, lung, ovary and prostrate. This analysis, published in *Preventive Medicine*, has estimated the health cost of USA's current omnivorous diet at $ 28 billion to $ 61 billion a year. Its authors say, 'the combined medical costs attributable to smoking and meat consumption exceed the predicted costs of providing health coverage for all currently uninsured Americans'. Further they said, 'The health effects of an omnivorous diet may result from the presence of meat, the displacement of plant foods, or both'. In an accompanying editorial, Dr Randall White and Suzanne Havala have written, 'The cost of American way of eating has become unsustainable'.

Unfortunately, almost all these concerns are being neglected and existing distortions are getting worse due to the domination of quick profit interest of agribusiness in policy matters while ignoring the wider, longer-term and much more significant interests of ordinary farmers and consumers. A lot of the confusion arises from treating agriculture much like an industry (or even mining) with a single minded pursuit of maximising short turn production and profit. The fallacy of this widely held view has been very aptly exposed by Wendell Berry. 'The farmer differs from the industrialist in that the farmer is necessarily a nurturer, a preserver of the health of creatures'.

> The economy of industry is, typically, extractive. It takes, makes, uses, and discards, it progresses, that is from exhaustion to pollution. Agriculture, on the other hand, rightly belongs to a replenishing economy, which takes, makes, uses, and returns—it involves the return to the source, not just of fertility or of so-called wastes, but also of care and affection.

Arguing why this basic understanding of agriculture is so widely missed these days Wendell Berry writes, 'The "free market"—the unbridled play of economic forces—is bad for agriculture because it is unable to assign a value to things that are necessary to agriculture. It gives a value to agricultural products, but it cannot give a value to the sources of those products in the topsoil, the ecosystem, the farm, the farm family, or the farm community. Indeed, people who look at farming from the standpoint of the "free market" do not understand the relation of product to source. They believe that the relation is merely mechanical because they believe that agriculture is or can be an industry. And the "free market" is helpless to suggest otherwise'.

There are thus very strong reasons why several economists, other allied experts and politicians advised by them are unlikely to see and appreciate the true nature of agricultural work. This also explains why governments are so reluctant to implement the really needed agricultural reforms, even though many experimental initiatives of farmers are providing a strong case for much needed reform.

Several of these initiatives have been mainly associated with organic farming of various types, with several well-recorded cases of how productivity can be maintained or even enhanced while protecting the soil and avoiding chemicals. Other experiments go beyond this into several cooperative and community efforts which bring out many of those dimensions of agriculture which are missing in the industry-oriented understanding (or rather misunderstanding) of agriculture. Trauger Groh, one of the pioneers of this movement in the USA, has mentioned some of the important aspects of community supported farms in his book, *Farms of Tomorrow:*

> Some things are typical for all community supported farms. In all of them there is a strong dedication to quality; most of them are organic or biodynamic farms, most of them show great diversification, most are integrated farm organisms having their own livestock and thus their own source of manure, or they are aiming in this direction. At all of them, far more people are working regularly per 100 acres than in conventionally run farms; and generally there are just many more people around participating in all of the dimensions of agricultural life: working, relaxing, storing, shopping, celebrating. This human element is of enormous importance. It shows that these farms have something to offer beyond good food. They embody educational and cultural elements that draw the interest of many people.

Important as these initiatives are for choosing alternative paths for agriculture, these need to be backed by wider action to counter the current international trends of dominance of food and agriculture by some giant agribusiness interests. A wide network of farmers, consumers, environment, safety, animal rights and other activists will have to make a concerted effort to stand up against these forces, giving more and more people a chance to understand how the all-important food and agriculture systems have been destroyed by such ruthlessly commercial interests and how the situation can be saved by environment friendly initiatives of farming communities in cooperation with other concerned people.

Towards a Harmonious Relationship
Forests, Wild Life and People

Recent debates in Washington have revealed a latent national guilt at having all but destroyed the great woodlands of America in less than 200 years. The axing of the forests turned the timber industry into one of America's largest and most profitable. First to go were the hardwoods and white pine, then the cypress groves of the south-west, then the pine woods of the great lakes. The American and Canadian Pacific forests were the last to feel the axe. Over the past 40 years, however, all but some 30 per cent has gone. Most of what remains is in Alaska, making the state the last resort for those who want to avoid the 'sins of the past'. But ultimately destruction also came to the Tongass Forests of Alaska, and that too in a cruel way.—Peter Stothard

THERE HAS been growing concern over the continuing fast pace of destruction of natural forests at a time when it is more important than ever before to protect the remaining natural forests. In many countries what has been practised in the name of modern and scientific practices has in fact been completely unscientific as this has provided justification for large-scale destruction of natural forests.

As an example of the single-minded, one track attitude of looking at nature (forests in this example) only from the point of view of extracting quick profits from it, some documents prepared by topmost forest officials in India can be quoted.

The document of India's fifth five-year plan said clearly, 'The primary object of the plan is to initiate measures for increasing production of industrial wood and other forest products by a changeover from conservation-oriented forestry to a dynamic programme of production forestry, aiming at clear felling and creating large-scale man-made forests with the help of institutional financing'. A little later, the National Commission on Agriculture went to the extent of saying: 'Production of industrial wood will have to be the raison d'etre for the existence of forests'. The officials and experts who prepared the draft of the five-year plan of the forestry department of Uttar Pradesh (1974-79) revealed that they were very receptive to the wider industrial demand for tree-pulp, as they wrote: 'There is greater demand for species which can be sliced or peeled or pulped for fabrication into a very wide range of products for consumers. Traditional hard timbers like sal need to be gradually replaced by species which are softer and faster growing'.

As long as such attitudes remain, not much can be expected from much talked about ambitious plans to protect environment. What is more, such distorted polices have created entirely avoidable conflicts between the livelihood of those who rely on forests and the protection of forests and wildlife. In fact wildlife protection effort based on fenced-off areas approach has alienated precisely those forest communities which can protect the habitat of wildlife. Tourism centered considerations have also fed this distortion and discouraged a balanced approach to protecting all forms of wildlife along with the livelihood of people living in and around forests. The bias in favour of the more majestic and colourful animals at the cost of less attractive and glamorous ones distorts the basic considerations (of reducing pain and ecological disruption) and due to this distortion it becomes difficult to secure the involvement of a larger number of people.

A related distortion is the bias towards the relatively small protected area of parks and sanctuaries to the relative neglect of the importance of animal rights in other areas including ordinary forests, rural areas and to some extent even urban areas. If all types of animal life is considered together then many more of them live outside protected areas than within protected areas. Their survival is influenced by many factors such as forest and agricultural policies, the extent to which pesticides and other poisonous chemicals are used in agricultural fields, what kind of dams are built on rivers, to what extent rivers and oceans get polluted, how destructive fishing practices are used, are the restrictions on wildlife product trade strictly enforced or not, whether restrictions (voluntary or legal) exist against the indiscriminate killing of animals etc. Breeding and disease control policies followed in the case of domesticated animals are also very important because wrong policies can easily lead to the replacement of indigenous, hardy, disease-resistant breeds with new breeds not really suited to local conditions and more vulnerable to several diseases.

Thus what is really needed is a concern for other forms of life everywhere on earth, and giving their rights and their problems a place in all relevant aspects of our development planning—whether this relates to agriculture, animal husbandry, forestry, water pollution, dams, agrochemicals etc. The overemphasis on a small area of sanctuaries and parks comes in the way of this wider perception of the problem and this leads to a false sense of our responsibility being over with the creation of protected areas.

Another serious distortion which has created a lot of avoidable controversy is that some animal rights activists unfortunately give priority to those ways of working which bring them in direct conflict with human beings, that too generally needy and poor people who themselves need some help in their survival struggle. This is most clearly seen in the efforts to evict villagers from places where some protected areas are being established. Surely there must be some ways for protecting wildlife other than those which call for displacement of people, particularly keeping in view the fact that agricultural fields also attract many types of birds. Yet, somehow, this way of protecting wildlife has been preferred to the exclusion of others which can prosper on a protective relationship between villagers and other forms of life.

Another example of avoidable confrontation is that when some wrong practices on the part of pastoral people, fisherfolk or other such groups are noticed, some enthusiastic animal rights activists arrogantly blame such groups without pausing to think if these poor people are themselves the victims of wrong policies. Some of these communities have been victimised and alienated by development policies and programs of recent decades, and it was their crisis of survival that pushed them to some possibly destructive and cruel practices. The problem should be seen in this wider perspective so that the grievances of these people can also be taken up while persuading them to give up any practices that cause great pain and suffering to other forms of life.

After all, the aim is not to create controversies that do not take us anywhere but to actually achieve a reduction in the suffering caused to animals while also protecting the ecological balance. We cannot achieve this by antagonising those very groups of people who live nearest to wild animals, or whose own livelihood depends on other forms of life. Animal rights activists have to work with these people, not against them and this requires that genuine efforts to understand such groups and their problems should be made.

In many present-day parks and sanctuaries, while a handful of elites amuse themselves, ordinary villagers become increasingly alienated from the otherwise highly justified movement of protecting the rights of other forms of life. The task of protecting animals and birds has to be ultimately carried out with the help of ordinary people. Therefore, it is important to include the important task of protection of animals in the wider framework of social change in such a way that ordinary people see it as their moral duty to reduce this distress as much as possible. Once this attitude exists on a large scale, then countless ways of reducing the distress of other forms of life will be found by people themselves.

There is in fact a rich tradition in many communities respecting the rights of other forms of life, going in some cases to opposing any form of violence against them. Such traditions have been disrupted in some cases because of the reckless ways in which some mercenary interests have taken advantage of the poverty of people to pay them to collect frogs and butterflies or to hunt valuable species like musk deer. Instead, why can't forest dwellers be given modest salaries to collect information on the mercenary elements who are destroying the botanical and zoological wealth of forests? Why can't they be mobilised to collect information on the various varieties of birds and insects that exist in the forests after imparting them a little training for this work? A network of such trained barefoot zoologists can prove invaluable in knowing more about our wildlife and in protecting it. In this and other ways, the conflicts that appear to exist between protection of animals and livelihood of people can be removed as much as possible and instead, protection of animals can become the livelihood of many more people particularly forests dwellers.

The relationship between human beings and other forms of life is one of the most significant aspects of life on our planet. It also happens to be an aspect which has received very little attention.

Any evaluation of this relationship should be based on some well-established and self-evident facts.

- This planet clearly belongs to all living beings and so a broad principle should be the protection of all and, at the very least, the survival of all species.

- All living beings experience pain so another guiding principle should be to minimise this pain as much as possible.

- While this planet is meant for all living beings, human beings clearly have a unique role to play in the planet which cannot be equalled by any other species. Among all species, human beings alone can cause changes (such as climate change, ozone depletion or nuclear warfare) which can threaten all forms of life on this planet. On the other side, only human beings can intervene to protect endangered species. However mighty lions may be , they are not capable of calling international meetings or setting up the agenda for protecting endangered species. This role clearly belong to human beings alone.

- Given this special role of human beings as the most powerful, active and decisive species on earth, human beings should use this position not with arrogance but with the utmost compassion for protecting other species and minimizing their pain and suffering.
- However, like all species, human beings also obviously retain the right of self-defence. If a tiger attacks a villager or his cow, the villager obviously has the right to protect himself/herself and his/her farm animal, and if in the process the tiger is killed, the villager cannot be blamed for this.

If an insect threatens the health of a village or its crops, the people of this village certainly have the right to get rid of this insect. However, they should be careful to use as specific measures against this troublesome insect as possible, so that other forms of life are not destroyed or disturbed. A poisonous chemical spray with the power to kill many neighbouring innocent species should be avoided for several reasons, one of which is the concern for those tiny species that have not harmed us but will be killed without any reason by the chemicals.

In their relationship with other species, human beings should be as protective as possible while retaining the right of defending themselves against those species which attack human beings or threaten to harm human beings. This is the essential principle on which policies towards other species should be based. In the practical realities of the existing world, such policies will run into several problems and clash with several interests. In such situations a compromise will have to be reached while trying to remain as close to this principle as possible. While trying to make this principle a practical reality, we can learn a lot from the lifestyles of those communities which have tried to live in harmony with other species, particularly wild species.

The parks vs people controversy, or the controversy that has arisen from the disruption of life of villagers caused by the creation of wildlife parks, should be examined in this context.

Laws on wildlife parks differ from country to country, but in most places these affect the life and livelihood of local villagers very adversely. In India, the existing law requires people to be evicted from 'national park areas' while in 'sanctuary areas' their rights to collect forest produce essential for their life and livelihood are seriously curtailed.

Ashish Kothari, a social scientist who has specialised in studying the impact of protected areas, says, 'Traditional activities have suddenly become "illegal", and villagers report considerable harassment and bribe taking by forest staff'. To add insult to injury, the reasons for this curtailment of rights have rarely been explained to villagers and viable alternatives have rarely been provided....

However, all this appears mild compared to the much more alarming reports from some other countries. According to Panos briefing on 'People and Parks', around 2,000 people are alleged to have been killed and 30,000 driven from homes to create a wildlife reserve in Burma.

Ramachandra Guha, delivering the keynote address to the Conference on Wild life and Human Rights in Asia at the University of Oslo said,

> The belief in a total ban on human intervention is misguided. Studies show that the highest levels of biological diversity are often found in areas with some (though not excessive) human intervention. In opening up new niches to be occupied by insects, plants and birds,

> partially disturbed ecosystems can have a greater diversity than untouched areas.

According to David Western, in East Africa,

> ...the ending of human activity in the parks, such as fires and shifting cultivation, has reduced biodiversity. Those human activities created the patchiness of terrain that encouraged more species. By pulling out the human components, and maintaining too many elephants we are losing biodiversity.

Ashish Kothari writes,

> A considerable part of the influence on Indian wildlife policies and programmes comes from the West. The American-European model which in the late 19th and early 20th century was being applied in the colonial tracts of Africa was also imported into India. The essential feature of this was the 'closure' of natural habitats to human populations... It was completely lost on Indian conservationists and administrators that such a model was unsuitable to situations where there is considerable human settlements inside these areas, and a strong dependence on their natural resources.

In Bharatpur Park area of India, villagers protested when the authorities banned grazing. In the ensuing conflict seven villagers were killed. A report of the Bombay Natural History Society concluded that here buffalo grazing was 'an integral part of the ecosystem, helping to counter the tendency of the wetland to turn into a grassland'. In some other cases also, including the famous 'Valley of Flowers' in Himalayan region, bans an traditional grazing had a negative impact on diversity and required a reintroduction of grazing or grass cutting.

Keeping in view such research findings a question can be

raised whether the interests of both human population as well as wild species will not be better served by giving up the notion of parks as areas devoid of human intervention. A better, more rewarding approach will be to look at how local people can prove helpful in protection of animals, birds and plants. Once this approach is followed, it is amazing to learn how many skills are already available with local people.

According to the Panos Briefing on People and Parks, 'The Kuikuru people in the upper Xingu valley of the Amazon rain forest can differentiate between 262 kinds of trees and 600 kinds of animals, and have found lasting use for up to half of them as sources of food, medicines, cosmetics and other commodities'. According to Brazilian anthropologist Darrell Posey, ancestors of Kayapo Indians in the Amazon planted many 'forest islands' of fruit bearing trees within the forests, which they tended assiduously. Several communities of forest dwellers have been living in harmony with wildlife. They have self-imposed restrictions on forest-use aimed at maintaining the green cover. They also have self-imposed restrictions on hunting or causing any other harm to wild animals. There is much to learn from them regarding the way of peaceful coexistence with wildlife. Instead a policy has been followed to uproot them from park areas.

This should be given up and the budget meant for fencing off the park areas, policing it and compensating the displaced people should instead be used to provide various benefits and opportunities to local villagers so that they can become even more involved in the protection of biodiversity. They can be trained to become barefoot botanists and zoologists. They can help in gathering invaluable information about flora and fauna. They can be organised to chase away poachers and smugglers, or at least provide reliable information about their activities.

Not all forest communities provide good examples of peaceful coexistence with wildlife. But even in problem cases, probably the budget meant for fencing, policing and confrontations can be better used to create such opportunities as will encourage the people to move towards protective attitudes and identify their own economic growth with protection of wildlife.

According to ecologist R Sukumar, 'There is an urgent need to reorient management of our wildlife reserves so as to pass on economic benefits to local communities... if an adequate proportion of the income derived from tourism is retained by the local economy there would be increased motivation for people to value wildlife and their habitats'.

On the other hand, if this approach is not followed and local villagers are displaced from park areas, or their livelihood is gravely jeopardised then it is quite likely that some of them will be forced into working for poachers and smugglers. Wild life protection work which involves their displacement will no doubt evoke strong resentment among them, while at the same time making them desperate for income. In such a condition it will be easy for poachers to recruit cheap workers from among them. In fact, by displacing honest, hard-working villagers, existing wildlife laws are providing easy recruits for poachers.

Further, once villagers are removed from park areas, whatever check exists on activities of corrupt wildlife officials will be removed. There is no reason to expect that corruption levels among these officials are any less than or any more than what exists in the rest of society. In the complete absence of any human habitation, a handful of corrupt officials controlling a thousand sq km or so of rich forest and wildlife area can earn millions from illegal activities which harm forests and wildlife.

So, removal of villagers from park areas can be harmful not only for them but also for biodiversity. The approach should be to involve them in the protection of biodiversity. The great Indian ornithologist, Salim Ali, wrote,

> No conservation laws or measures can succeed fully unless they have the backing of informed public opinion, which in our case means the usually illiterate village cultivator. In other words, unless we can make the villager understand, and convince him [or her] of the logic in expecting him to preserve the tiger or leopard that has deprived him of maybe his sole worldly possession—the cow which moreover provided the meager sustenance for himself and his family—how can we induct his willing cooperation? Similarly, how can we expect him to see any sense in being asked not to destroy the deer or pig that have ravaged the crops which he has toiled for months to raise, and on which all his hopes are banked? Admittedly this is going to be a very difficult task, but I believe it is not impossible if we could but find the right approach. We have never really tried enough. Devising a realistic strategy is now a challenge to all conservationists.

The significance of this viewpoint—of working with people and not against them—can be seen very clearly in the example of Van Gujjars, a semi-nomadic pastoral community which was threatened with displacement from Rajaji National Park area in North India. For several years forest and wildlife officials continued to propagate van gujjars as being most harmful for forests and wildlife with single minded devotion, and on the basis of this propaganda they tried to justify their eviction from the park area. At this stage an NGO—Rural Litigation and Entitlement Kendra (RLEK) as well as some other concerned persons stepped in to help the van gujjars in protecting their traditional livelihood.

In addition, RLEK implemented a highly innovative program among the van gujjars, in which teachers not only lived with the nomads in their forest abodes (deras) but also travelled with them on annual Himalayan migration routes. This NGO helped them with several other development projects and participation in the democratic process which had been denied to them so far. But more than all this, both by its own field research and by media access, it enabled people and decision makers to obtain an entirely different view of the van gujjar community - a community which far from being an enemy of forests (as some officials had always insisted) had a well-established tradition of living in harmony with forests and wildlife. It now became known that the van gujjars had a long, self-imposed restriction on hunting and although they are Muslim they do not eat any meat except in a few cases goats on rare festive occasions. RLEK has also prepared a plan for Community Forest Management in Protected Areas (CFM-PA) to enable van gujjars, in cooperation with other villagers, to manage a part of the Park area.

It is interesting to see what this plan document, prepared on the basis of detailed, meticulous fieldwork, has to say about the relationship of van gujjars with forests and wildlife, a relationship which had been earlier described by most officials in terms of hostility (all quotes below are from the CFM-PA plan prepared by RLEK).

> Reputation builds slowly over the lifetime of the individual but it is easily lost through disreputable conduct such as lying, cheating, stealing, fighting, flirting with another's wife, or disrespect for elders. At first, because it seems almost inconceivable, the mistreatment of animals or the forest is seldom mentioned. And in fact many deny that

it could ever happen. However Van Gujjars are sure that if it did, it would bring immediate disgrace and like the other 'sins' mentioned above, would provide grounds for public condemnation by the Panchayat, possibly even beating and ostracism. By contrast, well-fed buffaloes and a healthy forest attract honor to a man. Nature, which is also God, smiles on such a one, trees bloom, the buffalo bring forth milk and the good man prospers....

All forest animals, whether fierce or gentle, are seen as interrelated components of the natural world and therefore to be cared for by mankind. 'They are our children,' it is said, 'for they have the same life as our own.' Moreover, 'buffaloes have us to care for them but just because wild animals have nobody to care for them does not mean they should be harmed.'

Jackals, snakes, leopards and even the irascible elephant are all given a respected place within the moral order of the natural world. Though different, 'like the five fingers of the hand', each has its part to play in the great scheme of things and no vengeance is taken when elephants destroy deras (as they occasionally do) or leopards make off with calves or goats, for such is the way of the natural world and one animal gives its life for another much as trees sacrifice their leaves to deer and buffalo. Moreover, when an animal's time has come, its death is but the will of God. Since they are regarded as kin, no creature of the forest may be killed and even injury to one is cause for personal disgrace and discipline before the elders. No forest creature may be eaten or skinned and if found dead should be buried, preferably with a prayer similar to that used for human or buffalo burials.

To walk through the forest with a van gujjar guide is a lesson in biodiversity; every species of tree is known, its quality as fodder, the timing of its leaf-fall, medicinal properties and so on. Every sound has meaning, every bird known and its habits noted, every fallen branch or tree noted.

People having such value systems are particularly well suited for community management of forests and protected areas. Several inspiring examples of forest people living in or near forests, protecting these forests have been reported.

About two decades ago in Advani forest of Tehri Garhwal district neighbouring villagers and in particular women had gone one step further. They had hugged the trees to prevent them from being axed. When the contractor sent his workers to cut some other trees, the village women also rushed to these trees to hug them. The police and the contractor had to be called off. The auction was cancelled. Kunwar Prasun an activist of Chipko movement closely involved in this effort says today, 'The saved forest provided great relief during the drought years and people from even far off villages were able to obtain some fodder here.'

In Pujar village of the same district, single-minded devotion of one villager, Visheshwar Saklani, and the afforestation effort continued by him over several years has led to the emergence of a beautiful forest containing about 50,000 trees.

A post-graduate student of social work from Delhi University, Pandurang Hegde, came to Tehri Garhwal district for field-level experience and he was very impressed by the movement to save forests. He went back to his home district of Uttar Kannada in Karnataka and started mobilising the villagers there for protecting forests. After a lot of mobilisation efforts people of several villages, like Saklana, drove away the contractors and saved the forests from being axed.

Pandurang Hegde wants to emphasise that the people's resolve to protect forests has continued to this day and this can be seen in several examples of good community management of forests by people in some villages of Uttar Kannada district. 'Unfortunately, we are going in for heavily aided projects of dubious merit while ignoring the potential of village communities to manage their own forests. In my recent travels I've seen very inspiring examples of successful village community management of forests in many places such as some villages of Shimoga district (Karnataka), Bolangir district (Orissa) and Hazaribagh district (Bihar),' Pandurang says. He emphasises that all of these are examples of spontaneous action by villagers. 'The initiative to protect forests has to come from village communities themselves.'

A Programme for Justice, Peace
Future Development and Environment Protection

ANY COMPREHENSIVE understanding of India's possible pathways in the near future should be linked to a wider understanding of the most essential features of present day world. India should of course plan its future in accordance with its needs, but this has to be seen in the wider context of the world's most serious issues and problems so that the path India chooses is in conformity with the world's most urgently felt needs.

The two most important features of the present day world are—(i) the deepening of a many-sided ecological crisis in critical ways so that it has now become a survival crisis threatening the life of countless species and endangering even humanity, and (ii) the continuation and even accentuation of many-sided strife made all the more dangerous in a situation of more and more destructive weapons including thousands of WMDs.

These two problems may appear separate but are related in significant ways, not the least because *the kind of international effort needed to resolve the first listed survival issues will simply not be possible in a situation of increasing strife and violence.*

In the context of this wider situation of the world, India must pioneer a path which, while solving India's basic problems of poverty and deprivation, should at the same time show the way forward for resolving the most threatening problems of the world. If India can achieve this, it will be a truly historic achievement which will win admiration all over the world, increasing respect for India and leading to demands for a wider role of India in world affairs without India asking for this. Above all, if India achieves this, then this will be a huge contribution to reduce distress in the world, now and in the future, bringing relief to hundreds of millions of people. Can India achieve this?

For such an achievement to be possible, India should try to establish broad consensus on three overwhelming priorities—(i) reduction of poverty and inequalities, ensuring basic needs of all, (ii) protection of environment with special emphasis on checking survival threatening problems such as climate change, water depletion and loss of biodiversity and (iii) peace and social harmony at all levels while ending all discrimination.

If the government, supported by enthusiastic people's participation and a broad consensus, can really ensure that the inter-related tasks become topmost priorities, then India can truly pioneer a brave new path which can solve India's most pressing problems, at the same time providing important contributions and lessons for resolving the most threatening world level problems.

With 2 per cent of the world's land, 1 per cent (or less) of oil and gas resources but 17 per cent of the world's population, India's quest for meeting the basic needs of its people and providing them satisfactory livelihood on a sustainable basis is a huge challenge. If this can be met while also protecting environment and keeping down GHG emissions to acceptable levels, this will truly be a very commendable achievement.

India is a nuclear weapons power which shares borders with two other nuclear weapon powers. India has been extremely prone to terrorist attacks from across its borders. It has already fought five wars with neighboring countries. It has coped with several secessionist movements and insurgencies. Crime rates, not always reflected in official data, are very high in many parts of the country. India has six major religions (plus many sects and adivasi religions), around 6,400 castes and 1,600 languages.

The much-discussed idea of India is that despite all these outward differences, all people can live without discrimination in India with security and equal opportunities. This is the basis for peace based on equality and justice (as distinguished from peace based on threats and submission). There needs to be an all encompassing commitment to peace based on justice, reflected as much in non-discrimination and equality at all levels in India as in attitudes towards neighbouring countries. Of course, national security and borders should be fully and firmly protected. But we should also remember that unity of country based on equality of all can also be our most powerful defence.

Of course governance reforms including significant reduction of corruption and crimes (and related criminalisation of politics) and improved transparency are essential preconditions for success of this agenda based on justice, equality, harmony and protection of environment.

But what has been happening in recent times is far removed from the real needs of the country. Today, we have a system of crony capitalism tied closely to inequality and injustice based globalization which is based on giving increasing concessions to big corporations, particularly some favoured ones, and also including some of the world's most infamous multinational companies. Highly favoured are those MNCs which are known to be very aggressive in trying to dominate food and seeds sector, using very hazardous technologies. To pave the way for such corporate led growth, the Planning Commission has been shut down arbitrarily. The environment is threatened more than ever before with aggressively marketed and ecologically destructive projects of big corporates. The corrupt are getting increasingly aggressive and violent, as the bleeding Vyapam scam and many mining scams show. Powerful criminals are killing witnesses, activists and journalists with hardly any fear to check them.

The existing development model is only a somewhat revised imitation of the earlier dubious short success of export led and big business led growth models. Even at the best of times these cannot go very far. The present times are particularly unsuitable due to overall low prospects of export growth to developed countries as well as to other big markets like China, enhanced by increasing protectionist tendencies on the part of the USA and the increasing likelihood of unrest in the Middle-East. In addition, the corruption and crime ridden polity makes it more difficult for this to succeed without scams and crises.

Even in the midst of such grim prospects, it is important to hold high the banner that another path exists—a path that can make India a pioneer in reducing poverty, inequalities, environment ruin and GHG emissions, while promoting peace and harmony at all levels.

While spelling out the future program in greater detail, we should not be misled by claims based merely on GNP growth. We should not forget that when in the recent past India's economic growth rate went up significantly for some years the majority of people continued to experience serious problems in meeting their basic needs and environmental problems continued to increase at a fast pace. Economic inequalities grew sharply and the struggle for survival for those at the bottom remained a grim one. Small farmers, several categories of artisans, self-employed people and workers faced increasingly difficult conditions in protecting their livelihood. In several ways, the basic structure of the economy weakened. There is thus a clear need for making important corrections and removing serious distortions before it is too late.

It is with this spirit that major policy issues need to be clearly discussed with an emphasis on what has gone wrong, what are the risks ahead and what corrections need to be made. We need mutually consistent policy options for various sectors, leading to an overall policy framework which can work in real life conditions and is also in conformity with the creation of a better world.

While preparing a policy framework to some extent we need change but to some extent we also need to recognise what is of value at present and we need to protect it.

Basic Economic Policies

The highest priority should be accorded to meeting the basic needs of people on a sustainable basis. To ensure sustainability, first of all the environment should be well-protected and natural resources should be used very carefully. Also the economy should have a sound base. The basic needs of people include the following— adequate availability of balanced food satisfying nutrition norms, clean drinking water, satisfactory availability of clothes and housing to ensure protection from weather extremes as well as dignity, access to education which opens up opportunities of progress as well as strengthens basic human values, access to means of protection of health, medicare and basic hygiene.

Livelihood of small and medium farmers, artisans, workers, other vulnerable employees and self-employed persons should be protected and linked more closely to meeting the basic needs of all people. Special skills should be well-protected.

Economic inequalities should be reduced significantly as a matter of policy with emphasis on improving the prospects of people in the lowest layers of economy while also preventing excessive concentration of economic power and policy influencing power in the hands of any person, group or corporate interest.

In several critical areas of the economy the public sector should continue to play an important role. The private sector obviously should also have an important role but subject to the condition that no industrialist or company can dominate the economy, its one or more important sectors, to acquire monopoly powers and interfere unduly in the functioning of democratic system. Corporate sector should be regulated carefully for responsibilities relating to environment, workers, consumers (or other end-users of products) and to the wider society. Multinational and foreign companies should be regulated very carefully. Cooperative sector should be reformed and strengthened to accept increasing responsibilities. Certain products and areas can be reserved for small-scale and cottage-scale entrepreneurs, cooperatives, small farmer production groups and self-help groups, particularly of women, with emphasis on meeting basic needs of villages and small towns as well as generation of more diverse livelihoods there.

Economic planning should retain an important role in ensuring the availability of goods and services which meet the basic needs of people, reducing inequalities, protecting livelihoods, keeping unemployment and inflation at low levels, providing essential infra-structure and avoiding foreign indebtedness. Planning Commission should be re-established with some important reforms to strengthen it and the process of five-year plans should be re-started.

In foreign trade the drift towards heavy imports of several non-essentials including gold should be avoided. Steps which reduce excessive dependence on imports should be emphasised, while the sovereign government's powers to reduce excessive or harmful imports should be reclaimed. Similarly, patent laws should be in line with national interest. Free trade agreements existing and proposed should be clearly examined to protect national interests with special emphasis on interests of vulnerable sections. India should play an important role in reform of the WTO, the World Bank and the IMF with the aim to make them more transparent and responsive to the real and economic justice-based needs of world. Steps should be taken in time to avoid heavy indebtedness, balance of payments problems and heavy dependence on uncertain 'hot money' inflows. The type of linkages due to which any wider international economic crisis affects us quickly and excessively unsettling our national economy should be avoided.

There should be a relentless campaign against the substantial 'black' part of the economy so that illegally held money can be recovered and used for constructive development tasks. This includes efforts to bring back black money deposited abroad using various secretive devices. This should be taken up in cooperation with other countries and organizations dedicated to this work. India should not merely rely on the lead provided by developed countries in this context but should instead seek to lead the developing countries.

Budgets should emphasise raising of adequate resources to meet the basic needs of all people. Luxury consumption and high profit areas should be taxed adequately, while the tendency to give heavy concessions and huge exemptions to corporate sector and rich sections has to be given up. There should be adequate budgeting for social sector and in addition the increasing privatisation and high profit orientation of critical social sector components like health and education should be checked.

Agriculture, Animal Husbandry, Forests, Fisheries and Rural Development

High priority should be given to rural areas. The distorted thinking which necessarily equates development with very rapid urbanisation and migration of displaced villagers to mega-cities should be rejected. Villages should be the main base of India's development. Even though land availability per family is declining with the passage of time, more diverse livelihoods can be provided in rural and semi-rural areas by encouraging village and cottage industries, including khadi, and protecting artisans' livelihoods. These can include traditional improved as well modern industries, subject to the condition that these are not destructive for environment and public health. Decentralized mixed renewal energy systems, for example, can be a new area of growth of rural employment. Field-level protection of traditional seeds and biodiversity can be another such area.

Land rights of all small and medium farmers should be well protected, while also providing encouragement for their mutual cooperation and the strengthening of rural communities. Their land should not be lost due to indebtedness or related distress conditions. Fertile agricultural land should as far as possible be saved for agriculture and chances of displacement of farmers should be minimised. Special care should be taken to protect the land rights of tribal communities and to ensure the proper protection of laws enacted for this purpose. If displacement cannot be avoided in some cases, then as far as possible the efforts should be to provide land in place of land. High priority should be given to make available cultivation land to as many landless farm workers as possible, using either provisions like land ceiling laws, or reclamation of new land using water conservation and other steps. Housing land with legal rights should be available to all rural households who are still deprived of it.

Ecologically protective low-cost, location-specific technology which seeks to make best use of local resources and conditions should be emphasised, an approach which includes organic farming, protection of traditional seeds and biodiversity, soil and water conservation, increasing green cover and forests. Organic farming should avoid the pitfall of avoidable expenses like costly certification and should be based on self-reliance and low costs, including mutual certification by farmers of each other's crops. Farmers' seed rights should be well-protected and seed-banks of traditional diverse seeds should be set up with the close involvement of farmers including elderly farmers and women. Increasing control of big companies including multinational companies or their subsidiaries over seeds and other critical areas should be checked. GM crops and the related technology should be strictly banned keeping in view their many-sided, serious and irreversible adverse impacts and hazards. Use of chemical pesticides and weedicides should be completely avoided or at least minimised as much as possible. Protection of various forms of life should be emphasised. All subsidies meant for agriculture should be given directly to farmers.

Water conservation as well as protection/regeneration of greenery provide the base for survival in the form of meeting basic needs of life and supporting basic rural livelihoods. Some existing provisions like rural employment guarantee can be strengthened for this. In terms of resource use, concentrating attention on smaller watershed programs as well as proper maintenance of existing canals will yield much better results instead of various new big and medium projects of dams and canals. Of course, this will also be ecologically much safer and will help to avoid a lot of displacement. Safety of existing dams should be a significant area of concern. Indigenous mixed tree plantation work which resembles natural forests should be emphasised and cutting of existing green trees should be minimised as much as possible. Indigenous trees which provide fodder and fruits should be emphasised in addition to trees with better soil and water conservation properties.

Protection and regeneration of natural mixed forests should get very high priority. The practice of raising monoculture plantations of commercial species of trees in place of natural forests should be given up forever. The role of forests in providing diverse forms of valuable foods and nutrition should be recognised and protected. Forest based livelihoods of tribal communities and other communities living in and around forests should be protected and promoted. Their co-operatives or groups should be the main beneficiaries of minor forest produce based sustainable livelihoods, which should emphasise also the protection of trees. These communities should never be displaced or evicted in the name of protection of wildlife and forest; instead, as far as possible, they should get livelihoods in this protection.

Animal husbandry should be encouraged with special emphasis on regeneration of pastures and fodder trees as well as protection of indigenous species of farm animals. Protection of indigenous breeds of cows and bullocks should get special attention. Fair price should be ensured to dairy farmers. Their cooperatives should be strengthened with special emphasis on the poor. Milk powder imports and oil cake exports should be discouraged. Pastoral groups particularly nomadic and semi-nomadic groups facing hard times should be given a helping hand.

Availability of essential food items in public distribution network should be linked to strengthening of small and medium farmers in all rural areas. All raw food items needed for public distribution system as well as various nutrition programs should be procured from local farmers at a fair price. As far as possible self-reliance in essential diverse food items at the local, district level should be ensured with internal trade filling in unavoidable gaps. In this, special care should be taken to ensure that the public distribution system with cheaper food availability is used to strengthen local farmers and does not weaken them in any way. Rules of WTO or any other international rules which stand in the way of strengthening local food or farming systems should be resisted. However, subsidies as well as overall costs will be automatically reduced once long-distance movement of grains, long-term storage and related food-loss are reduced considerably. In public distribution system, millets, pulses and oilseeds should also get a proper place.

Steps should be taken to free various kinds of produce from the grip of a few big traders and speculators so that farmers get justice and sudden escalations in price for consumers are also avoided. Direct contacts between farmers and city-based consumers for healthy, organic food can be encouraged by allocating space in specific city markets to clusters of villages, subject to certain conditions so that the weaker, smaller farmers can get more benefit.

Any obstruction by international agencies, WTO or others, in a well-organised system of food security and food self-reliance based on farmers' secure livelihoods and sustainable fulfillment of basic food needs of people, should be resisted strongly.

Industrial Development

The country should aim, to the extent possible and practical, for self-reliance to a significant extent in all essential consumer and capital goods.

While private, public and cooperative sectors all have important roles, domination by any industrialist or use of unfair means to surge ahead in one or more sectors should not be allowed. Public sector should be strengthened and reformed to fulfill its wider social responsibility while maintaining high standards of efficiency and entrepreneurial ability.

All industrial units need to abide by properly framed regulations related to pollution, environment, displacement, health, safety, workers' welfare, quality and consumers' concerns.

Special protective steps including reservation of certain items in production and procurement need to be provided for cottage and small scale units with special emphasis on khadi and handlooms.

Activities of multinational and foreign companies should be regulated carefully.

Banking and Insurance

The important role of nationalised or public sector banks and insurance companies (mainly Life Insurance Corporation of India) should continue. These should be reformed and strengthened to improve their efficiency, basic financial soundness and social responsibility and to minimise the possibilities of corruption and irregularities.. Private and foreign banks and insurance companies can have only a limited role, and should not be allowed to damage the special position of national institutions like the Life Insurance Corporation of India. The extremely serious problems of so-called 'non-performing assets' should be sorted out by ensuring due payment of enormous sums owed by influential borrowers. Without any fear or favour very strong actions should be taken against anyone found guilty of corrupt practices or colluding in them.

Infra structure, Energy, Minerals

Public sector companies should continue to have an important role in creation of strong and adequate infra-structure for development of the country. While infra-structure should be adequate, unnecessarily expensive and grand projects should be avoided. Care should be taken to minimise the problems relating to environment and displacement.

Special care should be taken to reconcile development and environment protection objectives in the area of energy, as the responsibility of keeping GHG emissions at low levels must be respected. For rural areas in particular decentralised mixed renewable energy systems can play an important role.

Mineral wealth should be used in the wider interests of people with special emphasis on the rights and welfare of communities living in mineral rich areas. Instead of trying to take out minerals as quickly a possible or maximising corporate profits, various middle level options and technologies which protect communities and environment should be explored and emphasised. Domination of decision making by corporate interests should be resisted strongly. Foreign and multinational companies should not have any leading role in development of minerals. Minor minerals should be extracted in consultation with gram sabhas, minimising any harm to environment, while mining mafias should be kept away and resisted to prevent any criminalisation of mining work.

Health

A strong foundation of good health can only be established by good nutrition and fulfillment of other basic needs. In addition, essential health services, medicines, vaccines and investigations should be accessible to all. Adequate budgetary provisions should be provided for this. To utilise this properly, tendencies of extracting very high and unethical profits in the supply of medicines and medicare (including investigations) should be strictly curbed, or else the higher budget can be gobbled up by the profiteers. Important changes in medicines policy are needed to make available all essential medicines at a fair price, with special emphasis on supply of generic medicines, while irrational medicines and vaccines should be discarded. The public sector should fulfill an important role in this. The government should accept the responsibility of health care, medicines and vaccines. As far as possible, all medicines should be provided free in primary health centres and all government hospitals, perhaps excluding the very rich patients.

Special medical courses designed to ensure adequate and satisfactory availability of doctors and para medical workers in rural areas should be taken up.

Several distortions in health budget spending need to be corrected so that integrated and balanced health planning can emerge which is linked to the real needs of the country and not to artificial priorities thrust upon by vested interests.

While indigenous medical practices should be encouraged, there is need to ensure rationality and standards so that undesirable trends (like mass marketing based on dubious rationality) can be checked.

Doctors and other medical personnel coming forward with the objective of serving poor people particularly in remote villages should get the necessary encouragement from the government. Irrational rules unfavourable to serving in real-life rural conditions should be changed.

Tendencies towards unjust patent laws, domination by multinational companies and excessive privatisation should be resisted at a wider level.

Education and Children

While emphasising right to education for all, the education budget should be increased significantly.. At the same time, the tendencies of rapid privatisation and extraction of high profits should be checked. Improvement of government schools should get the highest priority. Children of weakest and vulnerable households (like migrant workers and nomadic groups) should also be included with a system of evening schools/bridge courses and later integration with the mainstream.

Tendencies towards communalisation of education should be curbed. Instead, a secular approach to moral/ethical education should be introduced with emphasis on universal values such as equality of all human beings, rejection of all kinds of discrimination, compassion for all forms of life, honesty, hard-work and a spirit of service.

Cooperation and not competition should be emphasised, in studies, sports (team-spirit) and other activities. Health education including a firm message against all intoxicants and also emphasising importance of physical work should get due importance, Education should provide a balanced view of realities and real needs of the country.

Child labour and all forms of exploitation of children should be eliminated. Trafficking of children should be curbed strongly and missing children should be traced with a sense of urgency.

Trafficked and exploited children when rescued should be rehabilitated properly. Creative programmes for street children should be implemented and various homes for disadvantaged children should be improved.

Special care should be given to ending discrimination against the girl child and improving her education opportunities.

Youth should have adequate opportunities for livelihoods linked to creating a better world and they should be adequately informed about such opportunities of employment as well as self-employment. The creativity of youth should get more and more openings in villages as well as cities.

High education and research should be linked to the country's real needs and careful use should be made of scarce resources.

Science and Technology

Progress in science and technology should be linked closely to the country's real needs. Technical skills not only in institutions of higher learning but also in rural areas, in farms and workshops and factories should be recognised, encouraged and provided adequate avenues. Technology and engineering skills should not be narrowly linked to any vested interests but instead should be directed towards serving the country's high priority needs.

Old Age, Disability and Pensions

Senior citizens should have a place of respect and dignity and to facilitate this better social security, particularly pensions, are very necessary. Extensive pension reforms should be taken up to create a system of universal and adequate pensions.

All disability related discrimination should end. Adequate care should be given to meeting the special education, health and other needs of disabled persons, providing them access to all social places and facilities, apart from arranging adequate pensions. 'Disability' as well as 'old age' should be defined in a comprehensive way so that any deserving and needy person is not left out of rights like social security. Prevention of and early treatment for accidents, injuries and diseases likely to result in disabilities should be emphasised.

Society and Religion

All forms of discrimination based on caste, religion, gender, colour, ethnicity etc. should be curbed strictly in keeping with the constitutional precepts. Apart from implementing legal provisions this should also be taken up as public campaigns.

Continuing efforts should be made, and not just at the

time of tensions, to maintain communal harmony. Strict action should be taken against those responsible for spreading communal hatred and tensions.

Everyone has a well-established constitutional right to follow his or her religion, but definitely not to insult other religions. All religions are equal in the eyes of the State, and governments should carefully follow secular precepts avoiding any discrimination.

However, there should be adequate room for social reforms and narrow thinking should not stand in the way of changing or removing those customs or traditions which clearly harm society and cause distress. Social reform movements against child marriage, dowry system, discriminatory practices, liquor and intoxicants, pornography, superstitions, various exploitative practices under the influence of superstitions etc. should be encouraged.

Community and family ties at all levels should be strengthened and social cooperation for creative, philanthropic and reformist work should be encouraged.

Harmful practices in the celebration of festivals should be curbed by public campaigns and legislation where necessary. The effort should be to protect good traditions while fighting harmful ones.

A campaign against the increasing consumption of liquor and tobacco products in various forms as well as against drug addiction should be a very important component of the social reform effort. The increasing auctions of liquor shops in villages have to be checked to a very substantial extent.

Social reforms should seek to involve most sections of the community and, as far as possible, should avoid creating new conflicts.

Scheduled Castes, Tribal Communities, Denotified Tribes, Nomads and OBCs

The existing reservations should continue till real equality in all important respects is achieved. A big effort should be made to provide some land to the large number of dalit (or other) landless farm workers and provide other assistance to help them to emerge as small farmers cultivating their own land. The ban on manual scavenging must be backed by adequate rehabilitation opportunities. Artisan work relating to bamboo, leather etc. should be improved so that new opportunities emerge and better, cleaner work-conditions are available.

Land rights of tribals should be carefully protected and land acquired earlier illegally should be restored under the due process of law. The implementation of recent Forest Rights Act needs to be substantially improved and any possibilities of large-scale displacement should be checked. Rights of minor forest produce should be strengthened and new opportunities opened up in processing work. Livelihoods based on protecting forest and wildlife can be substantially expanded. PESA—Panchayats (Extension to Scheduled Areas) law for decentralisation should be implemented in the right spirit. However, separatist elements should not misuse this law.

Nomadic and semi-nomadic tribes and groups deserve sympathetic understanding. Options of improving their present life pattern and rehabilitation are open. Denotified tribes need to be helped to come out of various kinds of stigmas and problems, and new opportunities must be opened for them.

In all categories the most oppressed and neglected groups deserve special attention and help. Particularly among OBCs there is a need to be careful that the genuinely oppressed, left-out and neglected castes get more help.

Labour

Hard-won rights of workers and trade unions should be protected. Reforms should not be used as a pretext to reduce or undermine these rights. Occupational health and safety need much more attention. The rights of unorganised sector workers deserve more attention and funds. Social security of these workers should be ensured in a big way. Women workers deserve special attention in terms of protection and care. Existing laws which protect rights of workers should be implemented in the right spirit. Within the organised sector also the rights of contractual workers deserve more attention.

Women and Gender Issues

There is a clear need to provide equal opportunities to women and end all gender-based discrimination. A system of (at least) 33 per cent reservation for women in state and national legislation and 50 per cent reservation in the decentralised system should be in place in the near future. The ban on female foeticide and infanticide should be implemented strictly. Apart from providing essential facilities, special incentives should be offered to encourage girl students. Security of girls and women should get high priority and urgent steps have to be taken at several levels to ensure secure living and working conditions for women.

Various reasons due to which violence against women has continued to remain at high levels should be identified carefully and remedial action should be taken. The rise in sexual violence is a matter of deep concern and the causes should be carefully identified so that effective action can be initiated on the basis of proper identification. Rapid proliferation of liquor, other intoxicants and pornography needs to be checked. Deeply entrenched attitudes of society which when interacting with these recent trends can prove very dangerous also need to be identified and checked. Strong laws to protect women are needed, but at the same time any misuse of these laws should be checked.

Land and property rights should be jointly in the name of wife and husband, and the rights of single women should also be ensured, but inheritance laws should not be such as to transfer ownership of a village's cultivation land outside the village.

Efforts for a better understanding of women's perceptions of various development issues should be made.

No one should face discrimination on the basis of gender or sexual orientations.

Other Forms of Life

It is important to avoid an excessively human-centric view of life as life forms other than human beings also need care and compassion. Many of these are today endangered due to the absence of this care. Most of the life forms who have been farmers' friends—ranging from indigenous species of cattle and camels to earthworms and sparrows, are badly in need of protection today. Indigenous species of cows and bullocks deserve special care due to their many-sided utility, but other farm animals should not be neglected. Communities living near forests should get strong livelihood support in activities relating to protection of wildlife and their habitats. Similarly, communities like fisherfolk, boatmen, can be involved in the protection of fish habitats and protection of all aquatic life, while snake charmers can be useful in protection of reptiles due to their special knowledge. Use of chemical pesticides and weedicides should be minimised while GM crops should be banned. Stray dogs and other stray animals also need better care. Cruelty to animals in the name of laboratory experiments should be reduced as much as possible.

Protection of Environment, Checking Climate Change

Protection of environment is of the highest importance not only for preparing the base of sustainable development but increasingly for sheer survival of various life-forms including human beings. Protection of environment and reduction of pollution should get priority at all levels, including reduction of air and water pollution, soil and water conservation, protection of forests, reducing the spread of various toxic products and wastes etc. New forms of pollution such as threat from radiation of nuclear plants, or the threat from mobile phone towers, or the irreversible risk of genetic pollution should be given adequate importance in the environment protection agenda.

While a strong legal base is certainly needed for protecting environment, people's movements and their close involvement in environment protection are equally important. It is important to evolve environment policies which involve people instead of alienating them. Environment protection work should provide new livelihoods to people, instead of displacing them or taking away their livelihoods. A ban on destructive mining at any place, for example, should at the same time provide for protective and regenerative work at the site which will also provide employment. Special care should be taken for protection of eco-sensitive areas of special importance like those in the Himalayas or in coastal areas.

While the task of environment protection has always been important, its importance has greatly increased in times of extremely serious world-level threats like those of climate change. India, like all countries, needs to give adequate importance to reduction of greenhouse gas emissions as well as to adopt steps to cope with climate change related problems. These can become issues of topmost importance in the near future. Also, India should contribute adequately to world level justice-based efforts to check climate change which should persuade developed countries to contribute substantially to efforts to check climate change, accepting historical responsibility for high greenhouse gas emissions.

Protection from Disasters

Increasing harm from several natural disasters is already a serious concern, while the threat from disasters can increase substantially in times of more aggressive climate change. Therefore, according much higher priority to protection from natural disasters has to be a very important part of policy framework now and in the times to come. To be effective, this effort has to learn from past experience and be willing to correct serious mistakes made earlier, as is evident from the increasing damage even after vast amounts had been invested in flood protection. So both increasing budgets and correction of distorted policies are important for protection from disasters. Planning for reduction of loss of life in an earthquake, cyclone or tsunami should get very high priority, as also various measures to provide protection from prolonged or acute droughts.

Protection from Accidents

Damage and threats from transport (particularly road), worksite, and domestic accidents as well as new threats from high hazard projects have been increasing. Preventive and immediate response action can reduce the damage including loss of human life to a substantial extent. So a nationwide network of protection and quick response for all kinds of accidents in an integrated way should be created.

Displacement

Efforts should be made to reduce displacement at all levels as much as possible. At the level of policy formation it is necessary to keep in view the need to minimise displacement. To the extent displacement cannot be avoided, all efforts should be made for satisfactory rehabilitation including land in place of land and protection of community ties. Cases of those victims of displacement who suffered injustice in much earlier times should also be considered sympathetically so that they can get justice, even though belatedly.

Justice, Police and Crime

Long pending police reforms should not be delayed any longer. These reforms should be aimed at not only increasing the efficiency of the police but also their sensitivity and humanity. Dignity of policemen at lower levels should be protected.

Reducing crimes should have a multi-dimensional approach with special emphasis on reducing the social causes of crimes as well as breaking the nexus between crime and corruption and political power at higher levels.

Efforts to combat terrorism should be much better organised and all support-systems of terrorists, whether in the country or abroad, should be opposed and challenged on a continuing basis at various levels. Social conditions and grievances which fuel terrorism should be tackled effectively.

The justice system should give special attention to ensuring that innocent people are not implicated in crimes. Special efforts should be made to provide legal aid to the poor and needy, as also to help undertrials. Undertrials who have already served a jail sentence which is equal to the punishment of the offence for which they have been charged should be released.

The justice system is breaking down particularly in rural areas because of long pending cases and repeated visits to courts from long distances which only lead to further dates. Therefore, rural decentralisation should include some judicial provisions for settling disputes locally but with suitable safeguards.

Jails need extensive reforms to create more human conditions, with special provisions firmly in place for recognition and human treatment of political prisoners.

Housing and Homeless People

Ensuring legal rights to housing land to all households and improvements in housing programmes for weaker sections should get high priority in rural areas, as also meeting the special housing needs in disaster prone areas.

The housing needs of urban areas cannot be solved by high profit oriented builders. The government should accept the responsibility for large-scale construction of houses to meet the needs of the poor as well as middle class. The needs of homeless people should get priority and construction of adequate shelters should be ensured. Slums should not be evicted arbitrarily, instead improving various facilities in slums should be emphasised.

Political Parties, Governance and Corruption

Efforts to reduce significantly the role of big money, illegal 'black' money and criminals in elections and functioning of political parties should get high priority. Political parties should maintain complete records of income, expenditure and all donations which should be transparent and should be accessible under right to information law. Election expenses should be kept low, rules should be carefully followed but routine work should not be interrupted at election time.

Right to information should be protected and strengthened, with additional protection for those who expose corruption using RTI or in other ways and to 'whistle-blowers'. Anti-corruption laws and organisations need to be strengthened and improved, while new laws should be introduced in areas where needed.

An effective grievance redressal system which provides for time-barred responses, issues receipts for complaints received and fixes responsibilities (as well as penalties for non-action) should be in place as soon as possible.

Exemplary strong action should be taken once corruption allegations have been confirmed. Excessive protection provided to some sections of officials from anti-corruption action should be withdrawn but at the same time tendencies towards witch-hunt without significant evidence as well as motivated targeting of innocent persons should be checked.

Strong action should be taken particularly in cases of illegal transfer of money earned by corrupt practices outside the country, in tax havens or secret accounts.

Decentralisation and Panchayat Raj

Decentralisation should be strengthened in rural as well as urban areas. Gram sabhas and ward sabhas in rural areas (assemblies of all adult villagers) as well as equivalent units in urban areas should be strengthened as a base where people's real needs can be articulated, discussed and also documented. It is important to strengthen ward sabhas particularly in villages where gram sabhas can be too large to give everyone a proper hearing. (Or else panchayats can cover smaller areas.)

Some weaknesses of panchayats need to be corrected. The tendency of one or two persons to concentrate most powers of panchayat raj can be corrected by strengthening of gram sabhas, a more active role for panchayat samitis as well as for all elected ward members and possibly a rotation of the main head-person's post among all ward members. Decentralisation at the district level should be strengthened significantly so that the concept of a district level government can emerge which is much closer to the day-to-day problems and livelihood issues of people.

Decentralisation needs to progress rapidly within the basic constitutional principles of equality and non-discrimination, integrity and unity of the country, secularism and socialism (interpreted in this context as reduction of inequalities and ensuring justice to weaker sections).

Urban Development

Urban life should emphasise environment-friendly and secure conditions for all sections of people, with special emphasis on reduction of pollution, access of satisfactory housing to weaker sections as well as middle class, and significant reduction of crime (particularly crime against women and children).

Instead of high concentration of population in big cities, balanced development of smaller towns including kasbas or semi-urban settlements close to rural areas should be prioritised. Satisfactory essential facilities should be provided in all these settlements.

Scarce resources should be used carefully to provide essentials to all people instead of squandering them on grand and expensive projects.

National Security and Unity

National security is based not just on armies and weapons but even more so on the unity and determination of people to defend national interests. Hence, continuing efforts should be made to strengthen unity of people at all levels. Special attention should be given to justice-based unity in border areas, with a willingness to provide all democratic rights (except perhaps for very small periods of special security concerns) and an effective system for redressal of all grievances. Special care should also be taken to ensure such community participation that any victimisation of innocent persons at the time of security operations can be avoided.

Apart from overall improvement of anti-terrorism operations, efforts to break their higher-level linkages whether external or internal should be emphasised.

A high level of preparedness to defend national borders should be maintained, while at the same time improving negotiations with neighboring countries to resolve border issues and ease tensions. Corruption and commissions in arms purchase should be strictly curbed, while self-reliance and indigenous R and D should be strengthened.

Priority should be given to solving most expensive border disputes like Siachen glacier. The Kashmir issue can be resolved by strengthening democratic processes on both sides of the border, initiating dialogues and exchange, gradually opening up borders and recognition of these as the legitimate borders by all sides. But all this is possible only if disruptive forces of violence can be restrained.

Foreign Policy and International Affairs

There should be a deep commitment to friendly relations and peace with all neighbouring countries without sacrificing national security interests and protection of our borders.

The world is passing through extremely difficult times with the growing threat of irreversible climate change and life-endangering conditions on the one hand, and the stockpiling and proliferation of weapons of mass destruction on the other. India should play a leading role of responsibility so that effective solutions for these life-threatening problems can get the highest priority and effective, justice-based solutions can emerge before it is too late.

Narrow viewpoints of developed countries on these issues should be challenged by a broader unity of developing and least developed countries. In addition, the tendency of developed countries to go on trying to extend their special privileged position and dominate world economy and trade should be challenged and resisted. International trade and finance institutions should be reformed substantially, or what may be more practical ultimately, new international trade and finance institutions should be created which are more suitable for creating a new international economic order based on justice and equality.

India should participate actively, inside and outside the United Nations, for justice based peace and minimising the possibilities of war and internal strife. Unity and cooperation of all countries should be established to curb strongly all forms of terrorism and its causes as well.

India should support disarmament with renewed vigour so that substantial savings from arms expenditure can be diverted to development needs.

Minorities

Security and equal opportunities of all minorities should be protected and promoted. Communal harmony and national unity should be promoted actively on a continuing basis so that such conditions are created which minimise possibilities of communal violence. Those guilty of obstructing this path of peace, goodwill and security should face strict action.

Although minorities are generally taken to be religious minorities, reasons for being identified on other basis (such as region or caste) exist and any such effort of violence against any kind of minority should face strict action and all such tendencies should be curbed.

Sexual minorities should also be protected from injustice, discrimination and stigma.

Social Activists and Organisations

Social activists who seek to help weaker sections, oppressed people and reform society in other creative ways should get encouragement and protection from governments. Any efforts to harass or repress such efforts of social activists and organisations should be checked and protection should be available against such efforts.

Media

Media should be close to the realities and needs of the country, and should be in tune with the basic precepts and values enshrined in our constitution—equality, secularism, special concern for the weaker sections, national integration and unity. Freedom of media is very important and this is also a basic constitutional precept, but it should not be misrepresented to justify increasing corporate control to an extent that media is alienated from genuine concerns of the people and instead promotes narrow interests and viewpoints. Cooperative efforts of journalists should be encouraged. Along with freedom, socially responsible behavior of new social media is important and new technologies should not be misused.

Transport and Tourism

Importance of railways and public transport, keeping in view the needs of ordinary travellers, should get high priority. Roads should be safe and in good condition, but overspending of scarce resources on non-essential expansion and widening should be discouraged. Safety in all forms of transport should get very high priority.

Safety, hygiene and essential facilities at all places of huge gatherings including pilgrimages, festivals and fairs etc. should be emphasised. Economy tourism and safety of tourists should get high priority. Tourism should be linked to the better livelihoods of ordinary people.

Culture, Art and Literature

Rich cultural assets and folk arts in various communities should be promoted and encouraged, as well as protected from the onslaught of corporate controlled media, 'cultural imperialism' and pornography. Highly deserving but neglected talent among ordinary folk should be encouraged and helped to realise its full potential. Special efforts should be made to protect endangered languages. Efforts to bring cinema, theatre and literature closer to the real concerns and aspirations of ordinary people, as well as pressing needs of society, should be encouraged.

The Non-violent Way of Life

Gandhi Even More Relevant Today

THE GREAT contribution which Mahatma Gandhi made in the 20th century is well known. The struggle against colonialism was a very important feature of a greater part of this century. It was of the greatest importance from the perspective of justice and democracy. Mahatma Gandhi not only made a very important contribution to this struggle against colonialism but in addition blazed a new trail by insisting with all his moral force that all such struggles for justice and truth should be based on non-violence. This proved to be an inspiration for many other struggles against injustice in various parts of the world including those led by Martin Luther King and Nelson Mandela.

The other most defining feature of the 20th century related to the two world wars. As the world belatedly realised the importance of avoiding such disasters in future the importance of Gandhiji's insistence of non-violence as one of the most basic precepts of life, as a way of life, became even more important.

It was also realised that by perceptive observers that the roots of colonialism as well world wars can be traced to never-ending greed and the conflicts created by this greed. Here again the insistence by Gandhiji on simplicity and voluntary frugality as a way of life and as one of the basic principles of life tied up very neatly with his commitment to non-violence.

Unlike other political leaders of the world, he never tried to promise more and more to the people of his country but instead devoted himself to spreading the principles of non-violence, simplicity and limited needs as a desirable way of life. He moved the economic debate beyond the narrow and familiar capitalism versus communism debate and instead concentrated on basic principles like simplicity, justice, prioritising the needs of the poorest, decentralisation, creation of local self-reliant economies and villages to the extent it is possible/practical/useful, respecting labour, protecting sustainable livelihoods while discouraging avoidable luxuries and parasitic pursuits.

The danger of jobless growth does not exist in the Gandhian scheme of things with its insistence of technologies being in tune with the needs of protection of local livelihoods and skills instead of following the ruthless logic of unencumbered capitalist growth. Gandhiji may not have used the words environment protection but the kind of economy he visualised is the one in which the base of environment protection can be created.

If the root cause of colonialism (or neo-colonialism and continuing imperialist ambitions) and wars lies in greed, then it follows that to avoid these we need an economy and a way of life based on limited needs, voluntary and happy acceptance of frugality and commitment to non-violence. Thus, in the Gandhian vision we have a framework in which a wide base for realising the most important objectives of peace and environment protection can be realised while at the same time meeting the essential needs of all people in sustainable ways.

Mahatma Gandhi would have made many great contributions with these ideas and thinking in the emerging post-colonial and post World War II world if he had lived for another 15 years or so. However, his life was very tragically cut short by an extremely stupid and narrow-minded assassin and his coconspirators who had absolutely no understanding of how much the world, and not just the nation, needed the broad vision of Gandhiji in these critical years.

It would have been very interesting to see Gandhiji with his far-reaching ideas and tremendous moral force reaching out to the world troubled by the Cold War and the arms race. In independent India he would have had the support of his nation and its government in reaching out to the wider world which he did not have under the colonial government.

All that is a history of lost opportunities. But coming to the present times, the most defining feature of the 21st century is that there is nothing less than a survival crisis caused by a number of very serious environmental problems on the one hand and accumulation/proliferation of weapons of mass destruction on the other.

As the world gropes without much hope and with an increasing sense of futility for the entrenched leadership to find solutions for these pressing problems before it is too late, genuine hope can emerge if there are strong grassroots efforts for peace and environment protection.

To reach out to more and more people, these grassroots efforts should also be based on meeting the needs of all people on a sustainable basis. In this effort to reconcile the three objectives of peace, environment protection and justice based on satisfaction of needs of all people, the ideas of Gandhiji can be very useful, particularly as these are supported by the impressive record of Gandhiji in living a life based on these principles and leading several struggles and campaigns based on them.

Hence, while his contribution during the century of his life was great indeed, his contribution in the next century when the world is threatened by a man-made survival crisis can be even higher as a well-thought out application of his ideas to resolving the present-day serious existential crisis can be of very great value.

Joint Action from 7 billion

People's Role Critical in Resolving Survival Issues

THERE IS increasing agreement among senior scientists that several problems confronting our planet have now taken the form of survival or existential issues which threaten the very survival of many forms of life including humanity. The special life nurturing features of our bountiful and beautiful planet which led to the flourishing of millions of life-forms on it are threatened by man-made factors. The survival problems range from climate change to nuclear weapons, ocean acidification to robot weapons. About a dozen such problems can be identified easily.

Another critical aspect on which there is increasing agreement of learned scientists and policy analysts is that the response to resolving these survival or existential problems so far has been extremely inadequate and poor. When a crumbling building is in need of foundation repair, its irresponsible owners and managers may deny the information to the hapless inhabitants who continue with their easy-going life unaware of the life-threatening dangers lurking around them. Such is the condition of our once benevolent planet which is being pushed towards more and more hazardous conditions with each passing day.

The most glaring failures have been at the institutional level—both at the level of national and international institutions. While institutional reform at both these levels is essential for resolving survival issues, the recent experience has been that such reform is unlikely to occur on its own at the institutional level itself. Of course, some institutions can prove to be an exception and make a significant contribution. Exceptional persons from a larger number of institutions may also make a significant contribution on their own even now. But on the whole, the present state of institutional response cannot be expected, based on recent experience, to be adequate for resolving all or even some of the mutually related survival issues in time. In order to be able to fulfill their badly needed historic role of resolving survival issues before it is too late, these institutions need to be reformed or changed by a push provided from an outside source and this can only come from people and from movements and organisations of people. Herein lies the important, the critical role of ordinary people in resolving the most urgent survival and existential problems.

This may appear to be a rather distant dream today as at first glance most ordinary people appear to be unaware of the survival issues or to be alienated from them. The situation may be somewhat different in some countries with high education levels and higher environmental consciousness, but speaking for the entire world as a whole the sad fact remains that consciousness about most survival issues, their causes and implications, remedial actions needed for resolving them, relationships among survival issues and their links with most urgent concerns like justice and democracy is at a very low level just now. However, as the hope for resolving survival issues rests so critically on mobilisation of people, this is a challenge that simply has to be taken up on a big scale and with deep commitment. If sustained and sincere efforts in this direction are made, these are bound to pick up and, with the passage of time, the truth about these survival issues can almost certainly become the most important rallying point for people and people's movements.

This process of education and mobilisation will be much more successful and rapid if the solutions needed for survival issues are linked up in hundreds of creative ways with the solutions needed for resolving the day-to-day economic and social problems of people. Ultimately, all solutions at both levels have to be based on the basic principles of justice, equality, peace, environment protection and democracy. On the basis of strengthening these guiding principles at all levels, programs and plans can be prepared which integrate solutions of daily problems of people with the much broader remedial actions needed for resolving global survival issues.

In the process, the values and value systems which are most helpful to take forward this task will be discovered. These should be encouraged, their spread and sustainability should be helped in various ways. School and family are the most important places where the beautiful plant of these value systems should be planted so that children who grow up in this world of increasing survival problems have the values that are most needed to accept the exceptionally difficult challenges of their generation.

Perhaps the most important of these new values will be the value of world citizenship, a belief in the equality (at all levels) of all people who on this basis together take up the challenge of protecting earth, our one and only home. Such values while starting with childhood should be strengthened in youth. In addition, adequate avenues should be found for the creative energies of young people to be channelised along lines that are in keeping with what is needed for resolving survival issues, while also providing fulfilling livelihoods to them.

Such efforts on a large scale, from bottom to top, should provide the creative energy needed for bringing the necessary changes in national and international institutions so that effective action can be taken for resolving survival issues before it is too late.

Acknowledgements

My family for all the support,
Reena Mehta for her help and suggestions,
Kuldeep Kulshrestha for his many-sided help,
Ishwar Singh for helping me in many small ways,
My publishers, Renu Kaul Verma, Papri Sri Raman
and Team Vitasta for their kindness and support
AND
Special thanks to all the editors of newspapers
and journals who published parts of this book

VOICES

To think that the only way to promote economic growth in the so-called underdeveloped countries is to imitate as closely as possible the current practices of the advanced countries.... and to force every time-honoured institution into the service of material aims of purely western inspiration, betrays not only an astonishing lack of imagination but also a truly ominous lack of awareness of the dehumanizing deformities of the modern west.

—E F Schumacher

Ecology has caught us by the throat.

—Mikhail Gorbachev

Everywhere people ask: 'What can I actually do?' The answer is as simple as it is disconcerting: we can, each of us, work to put our own inner house in order. The guidance we need for this work......can still be found in the traditional wisdom of mankind

—E F Schumacher

Like the constant dripping of water that in turn wears away the hardest stone, this birth-to-death contact with dangerous chemicals may in the end prove disastrous

—Rachel Carson

If stressed beyond the limits of whatever happens to be the current regulatory apparatus, Gaia will jump to a new stable environment where many of the current range of species will be eliminated.

—James Lovelock.

We have to feel the heartbeats of the trees, because trees are living beings like us.

—Sunderlal Bahuguna, Chipko Movement

Citizen of an advanced industrialized nation consumes in six months the energy that has to last the citizen of a developing country his entire life.

—Maurice Strong

We are living in a world locked into a suicide pact, with deeply entrenched ways of thinking about energy and security which are designed, unwittingly, to keep us on course for disaster.... The ways out of the suicide pact are obvious, yet are deemed visionary by many people. We can only pray that the visionary becomes seen as the imperative before it is too late.

—Dr Jeremy Leggett

We are not going to do away with the great car economy

—Margaret Thatcher, 1990

I have never pondered such matters.... nor do I ever intend to.

—Enoch Powell, senior British politician, on environmental issues.

About the Author

Bharat Dogra's reports from many remote parts of India on rural distress and struggles have been wake-up calls for policy-makers and general polity to focus more on the marginalised sections of our society. He has written extensively on environment, development and welfare issues for over four decades. His articles and reports on these issues have been published in leading newspapers and journals in India and abroad. He has received several prestigious awards for his writings in English and Hindi.

On the basis of his wide-ranging experiences and research Bharat Dogra has increasingly argued in recent years that the environmental crisis is now taking the form of nothing less than a survival crisis and the next few years will be crucial for resolving this catastrophe. This book from the noted journalist and author calls for people's response as the only way forward to lessen this worldwide crisis.